Forest Landscape Restoration

Forest landscape restoration (FLR) is a planned process that aims to regain ecological integrity and enhance human wellbeing in deforested or degraded landscapes. The aim of this book is to explore options to better integrate the diverse dimensions – spatial, disciplinary, sectoral and scientific – of implementing FLR.

It demonstrates the value of an integrated and interdisciplinary approach to help implement FLR, focusing specifically on four issues: understanding the drivers of forest loss and degradation in the context of interdisciplinary responses for FLR; learning from related integrated approaches; governance issues related to FLR as an integrated process; and the management, creation and use of different sources of knowledge in FLR implementation. The emphasis is on recognizing the need to take human and institutional factors into consideration, as well as the more obvious biophysical factors. A key aim is to advance and accelerate the practice of FLR, given its importance, particularly in a world facing increasing environmental challenges, notably from climate change.

The first section of the book presents the issue from an analytical and problem-orientated viewpoint, while later sections focus on solutions. It will interest researchers and professionals in forestry, ecology, geography, environmental governance and landscape studies.

Stephanie Mansourian is an environmental consultant with a special interest in the restoration of forested landscapes and governance challenges related to large-scale forest restoration. For five years she managed WWF International's forest landscape restoration programme. As a consultant since 2004, she has been involved in a range of international environmental work, including studying lessons from past FLR work, analysis of the Sustainable Development Goals (SDGs), and assessing governance challenges of FLR, as well as broader environmental project development and evaluation work. She has a PhD in Geography from the University of Geneva, Switzerland, where she is also a research associate.

John Parrotta is the national programme leader for international science issues with the United States Department of Agriculture (USDA) Forest Service's Research & Development branch, and currently serves as a vice-president of the International Union of Forest Research Organizations (IUFRO). His research experience and publications include work related to tropical forest ecology, silviculture, forest restoration and traditional forest-related knowledge. He is a coordinating lead author of the 2018 Intergovernmental Science-Policy Platform on Biodiversity and Ecosystem Services (IPBES) thematic assessment report on land degradation and restoration.

The Earthscan Forest Library

Series Editorial Advisers:

John L. Innes
Professor and Dean, Faculty of Forestry, University of British Columbia, Canada

John Parrotta
Research Program Leader for International Science Issues, US Forest Service – Research & Development, Washington, DC, USA

Jeffrey Sayer
Professor, Faculty of Forestry, University of British Columbia, Canada

Carol J. Pierce Colfer
Senior Associate, Center for International Forestry Research (CIFOR) and *Visiting Scholar, Cornell University's Southeast Asia Program, USA*

This series brings together a wide collection of volumes addressing diverse aspects of forests and forestry and draws on a range of disciplinary perspectives. Titles cover the full range of forest science and include biology, ecology, biodiversity, restoration, management (including silviculture and timber production), geography and environment (including climate change), socio-economics, anthropology, policy, law and governance. The series aims to demonstrate the important role of forests in nature, in people's livelihoods and in contributing to broader sustainable development goals. It is aimed at undergraduate and postgraduate students, researchers, professionals, policymakers and concerned members of civil society.

Recent titles:

Forests and Globalization
Challenges and Opportunities for Sustainable Development
Edited by William Nikolakis and John Innes

Smallholders, Forest Management and Rural Development in the Amazon
Benno Pokorny

Managing Forests as Complex Adaptive Systems
Building Resilience to the Challenge of Global Change
Edited by Christian Messier, Klaus J. Puettmann and K. David Coates

Additional information on these and further titles can be found at www.routledge.com/books/series/ECTEFL

Forest Landscape Restoration
Integrated Approaches to Support Effective Implementation

Edited by Stephanie Mansourian and John Parrotta

LONDON AND NEW YORK

First published 2018
by Routledge
2 Park Square, Milton Park, Abingdon, Oxon OX14 4RN

and by Routledge
52 Vanderbilt Avenue, New York, NY 10017

First issued in paperback 2020

Routledge is an imprint of the Taylor & Francis Group, an informa business

British Library Cataloguing-in-Publication Data
A catalogue record for this book is available from the British Library

Library of Congress Cataloging-in-Publication Data
Names: Mansourian, Stephanie, editor. | Parrotta, John A., editor.
Title: Forest landscape restoration : integrated approaches to support effective implementation / edited by Stephanie Mansourian and John Parrotta.
Other titles: Forest landscape restoration (Routledge)
Description: Abingdon, Oxon ; New York, NY : Routledge, [2018] | Series: The Earthscan forest library | Includes bibliographical references and index.
Identifiers: LCCN 2018016548| ISBN 9781138084292 (hbk) | ISBN 9781315111872 (ebk)
Subjects: LCSH: Reforestation. | Forest landscape management. | Forests and forestry. | Forest ecology. | Restoration ecology.
Classification: LCC SD409 .F6135 2018 | DDC 634.9/56–dc23
LC record available at https://lccn.loc.gov/2018016548

ISBN 13: 978-0-367-58757-4 (pbk)
ISBN 13: 978-1-138-08429-2 (hbk)

Typeset in Sabon
by Wearset Ltd, Boldon, Tyne and Wear

We dedicate this book in loving memory to Suzy Mansourian and Paul Parrotta, who sadly left us in 2017 while we were working on this book. For so many years, their unwavering enthusiasm and support provided us with energy and confidence when we needed it most.

– The Editors

Contents

Contributors

Poorna Balaji is currently a PhD student at the Academy of Conservation Science and Sustainability Studies, Ashoka Trust for Research in Ecology and the Environment (ATREE), India. Her research explores the evolution and trajectory of conservation policies related to forest diversion for development, and its impact on livelihoods and local ecology.

Imogen Bellwood-Howard is currently at the Institute of Development Studies, Sussex, UK. She has a background in environmental science and agricultural geography, and is currently working on agricultural livelihoods and groundwater futures in East and West Africa.

Suhas Bhasme is currently a postdoctoral scholar at Ashoka Trust for Research in the Ecology and the Environment (ATREE), India. His research looks into the linkages between institutions, development and natural resource management, particularly in rural parts of India.

R. Patrick Bixler is an environmental social scientist and research fellow at the LBJ School of Public Affairs at The University of Texas at Austin, USA. He received a PhD in sociology from Colorado State University, USA. His current research explores conservation social networks, collaborative governance, and the role of non-profit organizations in shaping environmental practice and policy.

Agni Klintuni Boedhihartono is an anthropologist and artist who has worked with traditional communities in the tropics to help explore their pathways to sustainable futures. She has worked with forest-dwelling communities in the Congo Basin, Central America and most of all in her native Indonesia. Intu has worked for the United Nations Environment Programme (UNEP) and the International Union for Conservation of Nature (IUCN), and initiated and ran a master's programme in Development Practice at James Cook University in Cairns, Australia for eight years. She is now Associate Professor of Landscapes and Livelihoods at the University of British Columbia in Vancouver, Canada.

Rachel Carmenta is a Frank Jackson Research Fellow at Wolfson College at the Department of Geography and University of Cambridge Conservation Research Institute, UK. Her research focuses on the human dimensions of tropical forest conservation with a geographic focus in Indonesia and Brazil.

Renato Crouzeilles holds a PhD in Ecology and conducted his career at the Federal University of Rio de Janeiro, Brazil, for 12 years. He is interested in reconciling conservation biology, landscape ecology and forest landscape restoration with environmental management and public policies at multiple scales to be applied to conservation, restoration and adaptive management actions. In addition, he studies the effects of habitat reduction and fragmentation on biodiversity and ecological processes, and investigates global restoration patterns through meta-analysis. He is currently an associate researcher at the International Institute for Sustainability, a collaborator researcher at the Center for Conservation Science and Sustainability Rio (PUC-RJ) and a collaborator lecturer at the Graduate Program at the Federal University of Rio de Janeiro.

Iain Davidson-Hunt is a professor at the Natural Resource Institute, Clayton H. Riddell Faculty of Earth, Environment and Resources, Canada and a registered professional planner with the Canadian Institute of Planners. His main research focus at present is biocultural design as a practice to support innovation for small-scale food systems and territorial development. This work is undertaken through partnerships with indigenous peoples and local communities (IPLCs) in rural and remote regions of the Americas. He has also worked with IPLCs on territorial land use planning, ethnoecology, ethnobotany and biodiversity-based enterprise development.

Wil de Jong is currently a professor at Kyoto University, Japan. In recent years his research has focused on tropical forest governance, illegal logging, smallholder and community forestry, and forest transition and restoration. His over 140 peer-reviewed publications include multiple peer-reviewed journal articles, edited special issues of academic journals and monographs, and edited book volumes.

Christian P. Giardina has been a research ecologist with the USDA Forest Service for the past 20 years and has focused his work on climate change, invasive species, and fire impacts to forests and forest landscapes. For the past 10 years, he has been collaborating with native Hawaiian scholars in the exploration of how native Hawaiian knowledge, values and practices inform restoration. He addresses landscape restoration questions primarily in Hawaii, but also in the Federated States of Micronesia, the Republic of Palau, Fujian Province of China, and most recently the Darién Province of Panama and the Nilgiris District of Tamil Nadu, India.

Catarina C. Jakovac is a biologist from the University of São Paulo, Brazil, with a PhD in Production Ecology and Resource Conservation from Wageningen University, The Netherlands. She is a specialist in tropical forest ecology, and her research focuses on forest succession after natural or anthropogenic disturbances. As part of the 2ndFOR international network, she has collaborated in the understanding of secondary succession and the conservation value of human-modified landscapes at local, regional and continental scales. Catarina is currently an associate researcher at the International Institute for Sustainability, working on models for the prioritization of restoration and for estimates of carbon gain from restoration.

Theresa Jedd is an environmental policy specialist and postdoctoral research associate at the National Drought Mitigation Center at the University of Nebraska, Lincoln, USA. She received a PhD in political science from Colorado State University, Fort Collins, USA, and her current research seeks to understand vulnerability and adaptation in outdoor recreation under a variety of climate and weather conditions.

André B. Junqueira is a biologist at the University of São Paulo, Brazil, with an MSc in botany (National Institute of Amazonian Research) and a PhD in Production Ecology and Resource Conservation, Wageningen University, The Netherlands. He is interested and experienced in interdisciplinary research projects focusing on ethnoecology, socio-ecological systems, ecology of agroecosystems and historical ecology. André is currently an associate researcher at the International Institute for Sustainability, Rio de Janeiro, Brazil.

Frank K. Lake is a research ecologist for the USDA Forest Service-Pacific Southwest Research Station, Fire and Fuels Program. His research focuses on American Indian tribal and community forestry and natural/ cultural resource management strategies. His interdisciplinary methods address restoration ecology and the incorporation of traditional knowledge into wildland fire and forest management, and for evaluating climate impacts on cultural resources and tribal values. Frank received a BS from University of California-Davis, USA, in Integrated Ecology and Culture with a minor in Native American Studies, and his PhD from Oregon State University, Environmental Sciences Program, USA.

Agnieszka E. Latawiec is a co-founder of the International Institute for Sustainability in Rio de Janeiro (IIS-Rio), Brazil, and the executive director of the Institute. She coordinates the Centre for Conservation and Sustainability Science (CSRio) at the Pontifical Catholic University of Rio de Janeiro (PUC-Rio). She is also assistant professor at the Department of Geography and Environment at the Pontifical Catholic University of Rio de Janeiro, Brazil, and associate professor at the Faculty of Production and Power Engineering at the University of Agriculture in

Krakow, Poland. Agnieszka holds a BSc in Environmental Protection Engineering and an MSc in Environmental Protection, with specialization in soil and plant chemistry. She is interested in topics related to interdisciplinarity and broad aspects of land management.

Ning Li is an associate professor of Business Management, College of Economics and Management, Zhejiang Agriculture & Forestry University, China. Her work focuses on corporate strategic management, corporate sustainability and ecological economics. She studies how global forest industry companies address biodiversity and ecosystem services in supply chain management.

Melvin Lippe is currently at the Thünen Institute of International Forestry and Forest Economics in Hamburg, Germany. He has a background in tropical agriculture and environmental management, landscape modelling and remote sensing, and is currently working on land cover/land use change assessments and landscape scenario modelling in Ecuador, the Philippines and Zambia.

Jinlong Liu is a professor at the School of Agricultural Economics and Rural Development and the director of the Centre for Forest, Environmental and Resources Policy Study of Renmin University of China. He coordinates the International Union of Forest Research Organizations (IUFRO) Working Group on Traditional Forest Knowledge in Tropical and Subtropical Regions. His expertise includes participatory forest management, environmental policy, nature resource management and forest governance.

Veronica Maioli is a biologist with a Masters in Botany from the Federal University of Rio de Janeiro, Brazil and a PhD in Ecology from the State University of Rio de Janeiro. She is interested in multidisciplinary approaches related to the knowledge and conservation of Brazilian biological and cultural diversity. Her research involves socio-environmental systems, ethnoecology and landscape ecology. Additionally, she has been working for over 10 years as a consultant and analyst for environmental licensing and impact assessment for different companies in Brazil. Veronica is currently an associate researcher at the International Institute for Sustainability working with biodiversity valuation, private land owners' perception and ecological restoration in private areas.

Nitin D. Rai is a Fellow at the Ashoka Trust for Research in Ecology and the Environment, India. He uses a political ecology approach to understand the implications of state and market-based conservation policy for people and landscapes. Nitin is an editor of the journal *Conservation and Society*.

Jeffrey Sayer is an ecologist who has spent much of his career working to achieve conservation of tropical landscapes, especially forests. He has worked throughout the tropics and has also headed forest conservation programmes for IUCN and WWF. Jeff was founding director general of the Center for International Forestry Research (CIFOR) in Indonesia. Currently, Jeff works mainly in Eastern Indonesia on conservation and sustainable development initiatives. He is a professor in the Faculty of Forestry at the University of British Columbia in Vancouver.

Juliana Silveira dos Santos has worked with remote sensing, geographic information systems (GIS) and landscape ecology for 10 years. She has been applying these concepts and collaborating with the development of landscape tools to understand the impacts of land use changes in the natural ecosystems and to answer ecological questions. Juliana studied Agronomy at the State University of Rio Grande do Sul (UERGS), Brazil, before getting a Master's degree and a PhD in Remote Sensing at the Federal University of Rio Grande do Sul (UFRGS), Brazil and National Institute for Space Research (INPE), Brazil. Since September 2015, she has been a researcher and collaborator at the International Institute for Sustainability (IIS), Brazil, working on sustainable agricultural intensification while sparing land for nature and landscape planning, focusing on ecological restoration and biodiversity conservation.

Bernardo B. N. Strassburg is an economist with a PhD in environmental sciences from the University of East Anglia, UK. As an assistant professor at the Department of Geography and the Environment at the Pontifical Catholic University of Rio de Janeiro, Brazil, Bernardo's research is focused on understanding socio-ecological processes related to land use and in developing solutions for reconciling the conservation and restoration of natural habitats with food security and other land-related human demands. Bernardo is a lead author of the Intergovernmental science-policy Platform on Biodiversity and Ecosystem Services (IPBES) Global Assessment and has previously co-authored the IPBES Guide on the Value of Nature. Bernardo is also the founder and executive director of the International Institute for Sustainability (IIS).

Fernanda Tubenchlak is a biologist from the Federal University of Rio de Janeiro (UFRJ), Brazil. She studied part of her degree in Environmental Sciences at the University of East Anglia, UK, through the 'Science without Borders' programme. Fernanda is currently finishing her Master's degree in Ecology at UFRJ and is a research assistant at the International Institute for Sustainability. Her research focuses on agroforestry systems as a tool for forest landscape restoration, investigating the synergies and trade-offs between production and conservation.

Yadav Uprety received his PhD in Environmental Science from the University of Quebec, Canada. He works at the interface between social and natural sciences. His research focuses on restoration ecology, landscape dynamics, ecosystem services, human–environment interactions, biodiversity use and conservation, and traditional knowledge, among others. He has published several peer-reviewed journal articles on different fields encapsulating geographical regions from Nepal and Canada. Currently, Yadav is coordinating landscape programmes in Tribhuvan University, Nepal.

Andrea Flores Urushima researched urban and regional planning for 17 years. After receiving international awards and working with sustainable development plans in Brazil, she undertook research on regional network capacity at Kyoto University, Japan and taught ecocriticism at Kansai University, Japan. She is a member of the International Planning History Society and the French Culture Ministry Academic Network JAPARCHI.

Marieke van der Zon is currently an external PhD candidate with Wageningen University, The Netherlands, focusing on the impact of governance on conservation and REDD+ (reducing emissions from deforestation and forest degradation, and the role of conservation, sustainable management of forests, and enhancement of forest carbon stocks in developing countries). She has worked for over 10 years on socio-environmental themes in Latin America, Africa and Asia, including legal empowerment, rural energy, conservation, and environmental and social safeguards.

Kristina Van Dexter is a PhD candidate at the Department of Environmental Science and Policy of George Mason University, USA. Her doctoral dissertation focuses on agricultural land use change in post-conflict and frontier tropical forest landscapes, with a particular emphasis on agricultural production among small farmers. Her research engages a multi-sited (and multispecies) ethnographic approach, drawing on environmental science interfaces with feminist, postcolonial/decolonial studies, to study environmental governance across dynamic landscapes.

Bhaskar Vira is Professor of Political Economy at the Department of Geography, University of Cambridge, UK, and founding director of the University of Cambridge Conservation Research Institute. His research interests centre on the changing political economy of environment and development, especially in South Asia.

Ingrid Visseren-Hamakers is an associate professor at the Department of Environmental Science and Policy of George Mason University in Fairfax, Virginia, USA. Her research programme revolves around global environmental governance, with a specialization in international biodiversity governance. Her research is theoretically embedded in the political and policy sciences, while concentrating on 'integrative governance', defined as the theories and practices focused on the relationships between governance instruments and/or systems.

Carina Wyborn is an interdisciplinary social scientist with a background in human ecology. Her research focuses on knowledge co-production in climate adaptation and biodiversity conservation. Carina's research examines the connections between science, policy and practice, and the capacities that enable effective and ethical decision-making in conservation governance. In her postdoctoral research, Carina worked closely with US federal agencies in southern Colorado to develop approaches that integrate climate adaptation into land management decision-making in the context of uncertainty. Her doctoral research examined the relationship between science and governance in large landscape connectivity conservation in Australia and North America. She currently works as a research adviser for the Luc Hoffman Institute.

Anastasia Yang is currently at the Thünen Institute of International Forestry and Forest Economics in Hamburg, Germany. Her work focuses on the connection between interdisciplinary scientific analysis of forestry and social systems, particularly in the tropics, and policy and governance planning. Specializing in tropical forest systems, she has worked for the Center for International Forestry Research (CIFOR), focusing mostly on the South East Asian region.

Yeo-Chang Youn is a professor at Seoul National University, South Korea, responsible for teaching and research on ecological economics and forest policy. He also coordinates the IUFRO's Working Party for Traditional Forest Knowledge. He is an expert in the Intergovernmental Platform on Biodiversity and Ecosystem Services (IPBES), coordinating the Asia and Pacific Regional Assessments of Biodiversity and Ecosystem services.

Part I
Why integration matters

1 The need for integrated approaches to forest landscape restoration

Stephanie Mansourian and John Parrotta

Introduction

Humankind is modifying the planet at unprecedented rates: mass extinctions are happening before our eyes, and climate change is threatening our very existence (e.g. Steffen *et al.*, 2007). Forest and forested landscapes play a major role in the global carbon cycle (Le Quéré *et al.*, 2009; Pan *et al.*, 2011) by absorbing atmospheric CO_2 and other pollutants, regulate hydrological cycles, contribute to soil formation and erosion control, provide us with many goods such as medicinal plants, building materials and food, and harbour the majority of the world's biodiversity.

Yet we continue to lose forests every year, with estimates suggesting that around 230 million ha of forest have been lost over the 2000–2012 period (Hansen *et al.*, 2013). Other calculations indicate that we are losing approximately 15.3 billion trees per year (Crowther *et al.*, 2015). In addition to forest loss, the more subtle process of forest degradation (i.e. reduction in forests' capacity to produce ecosystem services as a result of anthropogenic and environmental changes) is a pervasive problem in many parts of the world (Hosonuma *et al.*, 2012; Kissinger *et al.*, 2012). Though more difficult to perceive and quantify than deforestation (Sasaki and Putz, 2009; Thompson *et al.*, 2013), forest degradation is estimated to have affected up to 850 million ha in tropical regions alone (ITTO, 2002).

At the same time, according to Hansen *et al.* (2013), we replanted 80 million ha between 2000 and 2012. Food and Agriculture Organization (FAO) figures suggest that between 2010 and 2015, annual forest cover increased 4.3 million ha in some regions (FAO, 2016a), much of it occurring in China through large-scale afforestation and reforestation programmes. Despite these recent gains, our current global forest cover stands at just below 4 billion ha, whereas it is estimated that 5,000 years ago it was around 5.8 billion hectares (FAO, 2016b).

In response to the high rates of forest loss and degradation, numerous tree planting activities and campaigns have been launched around the world in recent decades. Famously, the United Nations launched a 'Billion Trees Campaign' in 2006, and more recently, in 2017, three international conservation organizations (World Wide Fund for Nature (WWF) UK, the Wildlife Conservation Society (WCS) and Birdlife International) launched the 'Trillion Trees' campaign. A quick look at some headlines and websites highlights the scale of the effort: WeForest's website states that they planted 4 million trees in 2016 alone; the International Tree Foundation aims to plant 20 million trees in Kenya by 2024; the European Outdoor Conservation Association seeks to plant 2 million trees; the Enterprise Rent-A-Car Foundation is funding the planting of 50 million trees; the courier company DHL launched a campaign in 2017 to plant a million trees; and so on. Yet, one may legitimately ask: what is the purpose of all those trees? Which trees? Where are they planted? Why plant so many trees? And more importantly, how many trees survive beyond one year? Questioning the underlying purpose of planting trees leads to many ecological justifications: trees are planted to retain soil moisture (31% according to FAO, 2016a), to restore habitat for endangered species or to provide shelter for wildlife (13% for biodiversity conservation according to FAO, 2016a); trees are planted to store carbon, to recycle nutrients and water; and so on. However, decisions both to lose trees and to return them to landscapes are ultimately made by humans. Even in the case of forest recovery through natural regeneration, a decision is being made to set that

Figure 1.1 Forest landscape restoration can meet the needs of people and biodiversity. In Tanzania, elephants roam across the plains of Tarangire National Park, while people pay to view them and local economic opportunities are provided by ecotourism.

Source: photo © S. Mansourian.

land aside (e.g. from agricultural production or grazing) to regenerate and not to convert or manage it for other purposes, as regeneration occurs over a period of many years (Chazdon, 2008).

Understanding the causes of forest loss and degradation is the first step to restoring a forested landscape. Unless drivers of forest loss are well understood, and fundamental ones addressed, it will be difficult, if not impossible, to reverse this trend. Many restoration efforts have failed to be sustained because of underlying degradation drivers remaining in place. For example, perverse subsidies emanating from the agriculture sector put pressure on forests by encouraging their conversion in many parts of the world.

The major direct drivers of deforestation include agricultural expansion, infrastructure development, mining and urbanization (Geist and Lambin, 2002; Hosonuma *et al.*, 2012; FAO, 2016b). Of these, agricultural expansion has been the most important direct driver of forest loss, accounting for 80% of deforestation worldwide, the majority of which has occurred through conversion of tropical forests in recent decades (Gibbs *et al.*, 2010). Approximately two-thirds of deforestation in Latin America is linked to commercial agriculture, while in Africa and tropical and subtropical Asia subsistence farming is the major driver of land use change (Kissinger *et al.*, 2012).

Like deforestation, forest degradation is driven by a variety of forces, including unsustainable and illegal logging, over-harvesting of fuelwood and non-timber forest products (NTFPs), over-grazing, human-induced fires (or fire suppression in dry forests), poor management of shifting cultivation, and climate change (Chazdon, 2008; Hosonuma *et al.*, 2012; Malhi, 2012; Kissinger *et al.*, 2012). These drivers can be traced back to human pressures, notably population growth, land scarcity, urbanization and market forces, including rising global demand for specific products such as edible oils (Lambin and Meyfroidt, 2011; Kissinger *et al.*, 2012).

Addressing both the direct and the underlying causes of deforestation and forest degradation is typically problematic due to weak governance, inadequate policies, poor or inadequate cross-sectoral coordination, perverse incentives and illegal activities (Kissinger *et al.*, 2012). Further negative impacts on forests may result from a variety of policy failures including conflicting laws, unclear and/or overlapping jurisdictional responsibilities, or where responsibility for forests rests with the generally poorer and less powerful environment ministry. At the same time, conflicts between traditional (de facto) understandings, rights, ownerships and approaches related to forested landscapes and *de jure*, 'official' legal approaches to managing the same landscapes may lead to these landscapes being poorly managed.

Of course, these are generalizations, with many differences existing across the globe. For example, the forest transition curve (Mather, 1992; Kauppi *et al.*, 2006; Rudel *et al.*, 2005) serves to demonstrate that many

Western countries – but also a number of others in Asia, Africa and Latin America (e.g. China, India, the Philippines, Thailand, Cote d'Ivoire, Rwanda, Swaziland, Lesotho, Costa Rica, Cuba and Uruguay) are recovering forest after having hit a low in their forest cover (Hosonuma *et al.*, 2012).

While only a couple of decades ago, forest restoration was considered a specialized activity, undertaken by restoration ecologists, foresters and land managers with very specific objectives (increasing ecological integrity or improving timber supply), it is now an approach that is promoted to remedy numerous problems, from climate change to food insecurity or disasters. As it has escaped the exclusive domain of foresters and ecologists, so it has become more fuzzy and difficult to frame. Recently, there has been a widespread adoption of the term 'forest landscape restoration' as an attractive approach to tackle deforestation and forest degradation.

An integrated approach is particularly relevant to forest landscape restoration (FLR) given its dual dimensions of improving ecological integrity and human wellbeing. Reconciling different values and objectives in FLR is a major challenge and one that remains nearly two decades since the initiation of FLR work. The objective of this volume is to explore the opportunities that interdisciplinary and integrated approaches may offer to FLR implementation.

Specifically, this book seeks to fill a gap in knowledge related to FLR planning, implementation and monitoring by bringing together scientists from a range of disciplines and backgrounds to tackle the interdisciplinary dimension of FLR. Our intention is to expand the breadth and reach of restorationists so that they may tackle more effectively the challenge of restoring the millions of hectares of deforested and degraded forest landscapes around the globe. It stems from the recognition that drivers of forest loss and degradation are predominantly human-made. Further, many of these causes straddle scales (e.g. contradictions between local-level needs and international targets) and sectors (e.g. agricultural subsidies contributing to forest loss), and can be understood differently depending on individuals' backgrounds (e.g. a social scientist perceives the landscape differently from an ecologist). We contend that these compounding complexities may best be addressed through integrated and interdisciplinary approaches.

Through the diverse chapters in this book, we seek to answer three questions:

1   What are some of the integration challenges for FLR?
2   What can we learn from other large-scale land use initiatives, frameworks or approaches?
3   How can integrated approaches improve FLR decision-making processes?

We return to these three questions in our concluding chapter (13), using the body of knowledge generated by the contributions in this volume.

Why is an integrated approach to forest landscape restoration needed?

Given the underlying human dimensions of forest loss and degradation, as well as the multiple and cumulative factors involved, we maintain that more integrated approaches may help to improve our ability to successfully undertake large-scale forest restoration and thus achieve lofty global goals for restoration. In the face of continued forest loss and degradation, and a lack of significant progress on scaling up restoration that meets the needs of people and biodiversity, we see both a need and an opportunity to explore a different side of forest landscape restoration planning and implementation, one that considers the integrated nature of these challenges.

Foresters have long engaged in planting trees and managing forests for multiple goods and services. For decades, forest scientists and ecologists have sought to understand the dynamics of forest ecosystems, to restore ecosystem functions on degraded forest lands and improve habitat quality for key species (Hobbs and Norton, 1996; Higgs, 1997; Clewell *et al.*, 2004; Lamb *et al.*, 2005; Falk *et al.*, 2006). Many development organizations have seen the value of bringing back trees to enhance rural livelihoods, supply communities with fuelwood, improve water and soil quality, and protect agricultural fields and coastlines. Decision-makers and businesses are also increasingly aware of the financial benefits of using tree planting to protect water sources, prevent soil erosion, and capture carbon to offset greenhouse gas emissions. Yet, there is surprisingly little collaboration and joint work to ensure that multiple environmental, economic and social objectives of forest restoration can be fulfilled and somehow reconciled.

Scaling up local or site-specific efforts to restore degraded forest lands over larger areas is one means of considering multiple objectives. In the last couple of decades, 'forest landscape restoration' (FLR) has become a buzzword acquiring much visibility, notably in meetings of the three main multilateral environmental agreements, that is, the three Rio Conventions: the Convention on Biological Diversity (CBD), the United Nations Convention to Combat Desertification (UNCCD) and the United Nations Framework Convention on Climate Change (UNFCCC). Originally, it was defined by experts convened by WWF and IUCN in 2000 as 'a planned process that aims to regain ecological integrity and enhance human well-being in deforested or degraded landscapes' (WWF and IUCN, 2000). The definition emerged from a desire to ensure that the scale of restoration was sufficiently ambitious and able to meet both ecological and human objectives. Global commitments to restore millions of hectares under such banners as the Bonn Challenge on FLR, the New York Declaration on

Forests, the AFR 100 (African Forest Landscape Restoration Initiative) and the Latin America Initiative 20×20 have led to a rapid popularization of the term, if not necessarily effective approaches to accomplish these ambitious goals. In reality, the initial definition has been adapted by different actors to suit their purposes (Mansourian, 2018). Interpretations of the term and methods to implement FLR diverge depending on the viewpoint of policymakers, decision-makers, social scientists, natural scientists, foresters, farmers, land management organizations and environmental non-governmental organizations (NGOs).

Scholarly work on FLR has tended to emanate from the forest science community (e.g. Stanturf *et al.*, 2012a; 2012b; Lamb, 2014). Practical field-based projects have tended to be led by ecologists, restoration practitioners and environmental organizations (e.g. IUCN and WRI, 2014). Limited work on human dimensions of FLR or large-scale restoration has been published (e.g. Aronson *et al.*, 2010; Egan *et al.*, 2011). For example, most research to date evaluating the ecological success of restoration comes from North America and Australia, with little attention paid to socio-economic outcomes (Wortley *et al.*, 2013).

Expertise in restoration ecology has been applied to the restoration of relatively small-scale sites, focusing on reproducing ecological processes with limited attention to social dimensions of restoration (e.g. McDonald *et al.*, 2016). Instead, restoration ecologists have primarily focused on recreating systems that meet exclusively biophysical criteria, that is, authenticity, naturalness, structure, composition, function and dynamics (Palmer *et al.*, 2006; Chazdon, 2008; Burton and Macdonald, 2011; Stanturf *et al.*, 2014b; McDonald *et al.*, 2016). Further, until recently, little effort has been made to explore and integrate the knowledge and experience of local communities and indigenous peoples into restoration efforts (but see Long *et al.*, 2003; Joseph and Mansourian, 2005; Berkes and Davidson-Hunt, 2006; Egan *et al.*, 2011; Hill *et al.*, 2013; Martinez, 2014; Hessburg *et al.*, 2015; Lake *et al.*, 2017; UNESCO, n.d.).

Integrating different disciplines – and in some cases knowledge systems – presents a means of reconciling diverse perspectives and objectives. It serves to ground FLR in a wider set of realities than just those of foresters or ecologists. In this way, it may also be more relevant to individuals directly involved with or affected by FLR projects. For example, bringing the human dimensions of FLR forward and revealing at least some of their complexities helps project managers to better understand and address them. Specifically, it helps them to tackle the needs of rural communities more tangibly, rather than simply focusing on planting trees, which may be of little or no value to local people (e.g. in Vietnam, see McElwee, 2009 or in Rwanda, see van Oosten *et al.*, 2018).

Some key definitions

To clarify terminology used in this volume, in this section we review key definitional issues associated with FLR and the concept of integration.

Forest landscape restoration

Initially defined in 2000 as 'a planned process that aims to regain ecological integrity and enhance human wellbeing in deforested or degraded landscapes' (WWF and IUCN, 2000), FLR has seen a number of different definitions and interpretations since then as it has become much more widely used (see e.g. Sabogal *et al.*, 2015; Pistorius and Kiff, 2017). For the purposes of this book, we worked with the original definition of forest landscape restoration. Key elements in this definition are that FLR:

a   *is a long-term and planned process* – there is the intention to restore, and not just to label ad hoc natural regeneration as FLR (even if natural regeneration can be part of FLR), and a recognition that it is a process which takes time and will necessarily evolve in a dynamic way.

b   *aims to regain both ecological integrity and enhance human wellbeing* – there are two key dimensions to FLR: human and ecological. Ecological integrity can be understood as a measure of ecosystem function, composition and structure. Human wellbeing reflects many dimensions that include, among others, basic material needs, health, shelter, good social and community relationships, spiritual wellbeing, and personal security.

c   *takes place within a landscape* – the landscape represents a geographical space but also a means of integrating human and ecological dimensions. As a geographical scale, the landscape permits the definition of objectives that are more likely to integrate ecological as well as socio-economic dimensions (with different parts of the landscape performing different functions). As a means of reconciling both human and ecological objectives, the landscape enables trade-offs and negotiations.

The intention behind this definition of FLR is not to cover the entire landscape in forests, but rather, to improve the overall landscape functionality and to optimize the role of forests within that landscape.

Integrated, multidisciplinary and interdisciplinary approaches

Approaches that combine different disciplines can be integrated, interdisciplinary, multidisciplinary or transdisciplinary. Integrated approaches seek to bring different disciplines together. Multidisciplinary approaches aim to convene various disciplines to collaborate and approach a problem from their perspectives, but without changing their individual methods (Van den Besselaar and Heimeriks, 2001). Transdisciplinary approaches seek to unify the involved disciplines at 'the paradigmatic (metaphysical) level' (Adger *et al.*, 2003). More refined interdisciplinary approaches further seek to have different disciplines use one common methodology in their research (Van den Besselaar and Heimeriks, 2015). An interdisciplinary approach brings the added value of combining disciplines, strengths, tools and methods. Strictly speaking, an interdisciplinary approach creates its own common methods (Lele and Kurien, 2011; Van den Besselaar and Heimeriks, 2015). It can also hamper progress, as teams may require more time to agree on an approach (Massey *et al.*, 2006). We interpret interdisciplinarity as being a way of bringing together perspectives from different fields of study, and even different knowledge systems, in order to solve a common problem (adapted from McNeill *et al.*, 2001).

Our book presents a hybrid between multidisciplinary and interdisciplinary approaches. It is integrated from the point of view of the breadth of disciplines and expertise presented by our authors. The book is also interdisciplinary to the extent that we all adhered to the same overarching objectives and collaborated in attaining these objectives. Both multidisciplinarity and interdisciplinarity are reflected in the choice of editors and authors and the scope of the chapters comprising this volume. Nevertheless, we recognize the challenges and constraints of a truly interdisciplinary approach (Jones and MacDonald, 2007) and, like others (e.g. McNeill *et al.*, 2001), we cannot claim to have fully done justice to interdisciplinarity by adhering to one common framework and/or method. While this may be an imperfect attempt, we feel it nonetheless provides an important contribution to advancing the practice of FLR.

Structure of the book

Through this volume, we intend to demonstrate the obstacles encountered as a result of limited integration and interdisciplinarity and the consequent impacts on large-scale forest restoration and related programmes to date (Part I), as well as options, approaches and tools to improve interdisciplinarity (Part II). Governance and knowledge integration are highlighted as means for improving interdisciplinarity in Part III.

The first section of the book explores in more detail some of the challenges that have led to forest loss and degradation and also, maybe more importantly, to poor approaches being chosen for restoration that have met with only limited success or, in some cases, outright failure. It considers the lack of integration across sectors, scales, disciplines and knowledges (Chapter 2), and provides a categorization of the resulting silos. It also explores in more detail the important dimensions of different forms of knowledge and the lack of integration between Western science and Traditional knowledge (Chapter 3). The next chapter explores the social and economic inequalities generated by inappropriate land use decisions and their effects, particularly in the context of past restoration and reforestation efforts (Chapter 4).

Part II presents some of the frameworks that are currently being discussed in other land use contexts and considers their value and application specifically to FLR. Different approaches, systems and processes can help to better understand FLR as an integrated and interdisciplinary process. The aim of this section is to consider the opportunities, and constraints, for using existing integrated frameworks to support and accelerate FLR implementation. Our aim is to generate lessons from these related approaches for FLR implementation. These frameworks are social-ecological systems (SES), discussed in Chapter 5, landscape approaches (Chapter 6), land sparing/land sharing and 'Nature's Contributions to People' (Chapter 7) and agroecological approaches (Chapter 8).

The next section (Part III) focuses on governance and the integration of diverse stakeholders' perspectives, knowledge systems, objectives and needs. Governance – who takes decisions, how and under what conditions – is critical to the FLR process, particularly because the landscape dimension implies multiple stakeholders and consequently, multiple rights and ownerships. For FLR this is a particular challenge, notably because of the fact that land may have different values to different stakeholders, because of the scales (temporal and geographical) involved (and therefore the number of stakeholder groups), and because of ownership issues over not only land and forests, but also trees and other forest goods (including carbon, in the case of restoration for carbon sequestration). In the case of landscapes where indigenous and local communities reside, an added challenge – and opportunity – may be presented with respect to developing a shared vision of the nature and history of forest landscape degradation, and creating the enabling conditions to integrate diverse worldviews and knowledge systems into FLR planning and implementation. The aim of this section is to explore in greater detail the importance of governance and how it can facilitate or hinder FLR implementation. The chapters in this section are forward looking, seeking to highlight the available options (with examples and case studies) for ensuring the successful and long-term implementation of FLR. Specifically, we consider the role of different stakeholders in FLR and their relationships to each other (Chapter 9). Chapter 10 examines challenges associated with tenure and property rights and how they influence the effectiveness of FLR. This is followed by a consideration of the relevance of multi-level and polycentric governance to FLR implementation (Chapter 11). The final chapter in this section (Chapter 12) focuses on the importance of integrating knowledge in FLR implementation, critically examining the current emphasis on Western knowledge and the opportunities to incorporate Traditional knowledge held by typically marginalized stakeholders to support multidisciplinary approaches to restoration.

In the concluding chapter (Chapter 13), we draw on the preceding ones to synthesize key findings and implications for policy and practice. We reflect back on the three questions posed in this chapter. Our ultimate intention is to provide some added value to current literature and research on FLR while also providing useful findings to both policymakers and practitioners.

References

Aronson, J., Blignaut, J.N., Milton, S.J., Le Maitre, D., Esler, K.J., Limouzin, A., Fontaine, C., De Wit, M.P., Mugido, W., Prinsloo, P., Van Der Elst, L. and Lederer, N. (2010) 'Are socioeconomic benefits of restoration adequately quantified? A meta-analysis of recent papers (2000–2008) in *Restoration Ecology* and 12 other scientific journals', *Restoration Ecology*, vol 18, no 2, pp. 143–154.

Berkes, F. and Davidson-Hunt, I.J. (2006) 'Biodiversity, traditional management systems, and cultural landscapes: examples from the boreal forest of Canada', *International Social Science Journal*, vol 187, pp. 35–47.

Burton, P.J. and Macdonald, S.E. (2011) 'The restorative imperative: challenges, objectives and approaches to restoring naturalness in forests', *Silva Fennica*, vol 45, no 5, pp. 843–863.

Chazdon, R. (2008) 'Beyond deforestation: restoring forests and ecosystem services on degraded lands', *Science*, vol 320, pp. 1458–1460.

Clewell, A., Aronson, J. and Winterhalder, K. (2004) *The SER International Primer on Ecological Restoration*, SER, Washington, DC.

Crowther, T.W., Glick, H.B., Covey, K.R., Bettigole, C., Maynard, D.S., Thomas, S.M., Smith, J.R., Hintler, G., Duguid, M.C., Amatulli, G. and Tuanmu, M.N. (2015) 'Mapping tree density at a global scale', *Nature*, vol 525, no 7568, pp. 201–205.

Egan, D., Hjerpe, E.E. and Abrams, J., eds. (2011) *Human Dimensions of Ecological Restoration: Integrating Science, Nature, and Culture*, Island Press, Washington, DC.

Falk, D.A., Palmer, M.A. and Zedler, J.B., eds. (2006) *Foundations of Restoration Ecology*, Island Press, Washington, DC.

FAO (2016a) *Forest Resources Assessment*, FAO, Rome.

FAO (2016b) *State of the World's Forests 2016. Forests and Agriculture: Land-Use Challenges and Opportunities*, FAO, Rome.

Geist, H.J. and Lambin, E.F. (2002) 'Proximate causes and underlying driving forces of tropical deforestation: tropical forests are disappearing as the result of many pressures, both local and regional, acting in various combinations in different geographical locations', *BioScience*, vol 52, no 2, pp. 143–150.

Gibbs, H.K., Ruesch, A.S., Achard, F., Clayton, M.K., Holmgren, P., Ramankutty, N. and Foley, J.A. (2010) 'Tropical forests were the primary sources of new agricultural land in the 1980s and 1990s', *Proceedings of the National Academy of Sciences*, vol 107, no 38, pp. 16732–16737.

Hansen, M.C., Potapov, P.V., Moore, R., Hancher, M., Turubanova, S., Tyukavina, A., Thau, D., Stehman, S.V., Goetz, S.J., Loveland, T.R. and

Kommareddy, A. (2013) 'High-resolution global maps of 21st-century forest cover change', *Science*, vol 342, no 6160, pp. 850–853.

Hessburg, P.F., Churchill, D.J., Larson, A.J., Haugo, R.D., Miller, C., Spies, T.A., North, M.P., Povak, N.A., Belote, R.T., Singleton, P.H. and Gaines, W.L. (2015) 'Restoring fire-prone Inland Pacific landscapes: seven core principles', *Landscape Ecology*, vol 30, no 10, pp. 1805–1835.

Higgs, E.S. (1997) 'What is good ecological restoration?', *Conservation Biology*, vol 11, no 2, pp. 338–348.

Hill, R., Pert, P., Davies, J., Robinson, C.J., Walsh, F. and Falco-Mammone, F. (2013) *Indigenous Land Management in Australia: Extent, Scope, Diversity, Barriers and Success Factors*, CSIRO Ecosystem Sciences, Cairns.

Hobbs, R.J. and Norton, D.A. (1996) 'Towards a conceptual framework for restoration ecology', *Restoration Ecology*, vol 4, no 2, pp. 93–110.

Hosonuma, N., Herold, M., De Sy, V., De Fries, R.S., Brockhaus, M., Verchot, L., Angelsen, A. and Romijn, E. (2012) 'An assessment of deforestation and forest degradation drivers in developing countries', *Environmental Research Letters*, vol 7, no 4, p. 044009.

ITTO (2002) *ITTO Guidelines for the Restoration, Management and Rehabilitation of Degraded and Secondary Tropical Forests*, ITTO Policy Development Series No. 13, ITTO, Yokohama.

IUCN and WRI (2014), *A Guide to the Restoration Opportunities Assessment Methodology (ROAM)*, IUCN and WRI, Gland and Washington, DC.

Jones, P. and Macdonald, N. (2007) 'Getting it wrong first time: building an interdisciplinary research relationship', *Area*, vol 39, no 4, pp. 490–498.

Joseph, G. and Mansourian, S. (2005) Restoring landscapes for traditional cultural values. In: Mansourian, S., Vallauri, D. and Dudley, N. (eds), *Forest Restoration in Landscapes*, Springer, New York, pp. 233–238.

Kauppi, P.E., Ausubel, J.H., Fang, J., Mather, A.S., Sedjo, R.A. and Waggoner, P.E. (2006) 'Returning forests analyzed with the forest identity', *Proceedings of the National Academy of Sciences*, vol 103, no 46, pp. 17574–17579.

Kissinger, G., Herold, M. and De Sy, V. (2012) *Drivers of Deforestation and Forest Degradation: A Synthesis Report for REDD+ Policymakers*, Lexeme Consulting, Vancouver.

Lake, F.K., Wright, V., Morgan, P., McFadzen, M., McWethy, D. and Stevens-Rumann, C. (2017) 'Returning fire to the land: celebrating traditional knowledge and fire', *Journal of Forestry*, vol 115, no 5, pp. 343–353.

Lamb, D. (2014) *Large-Scale Forest Restoration*, Routledge, Abingdon.

Lamb, D., Erskine, P.D. and Parrotta, J.A. (2005) 'Restoration of degraded tropical forest landscapes', *Science*, vol 310, no 5754, pp. 1628–1632.

Lambin, E.F. and Meyfroidt, P. (2011) 'Global land use change, economic globalization, and the looming land scarcity', *Proceedings of the National Academy of Sciences*, vol 108, no 9, pp. 3465–3472.

Le Quéré, C., Raupach, M.R., Canadell, J.G., Marland, G., Bopp, L., Ciais, P., Conway, T.J., Doney, S.C., Feely, R.A., Foster, P. and Friedlingstein, P. (2009) 'Trends in the sources and sinks of carbon dioxide', *Nature Geoscience*, vol 2, pp. 831–836.

Lele, S. and Kurien, A. (2011) 'Interdisciplinary analysis of the environment: insights from tropical forest research', *Environmental Conservation*, vol 38, no 2, pp. 211–233.

Long, J., Tecle, A. and Burnette, B. (2003) 'Cultural foundations for ecological restoration on the White Mountain Apache Reservation', *Conservation Ecology*, vol 8, no 1, p. 4.

Malhi, Y. (2012) 'The productivity, metabolism and carbon cycle of tropical forest vegetation', *Journal of Ecology*, vol 100, pp. 65–75.

Mansourian, S. (2018) 'In the eye of the beholder: Reconciling Interpretations of Forest Landscape Restoration', *Land Degradation and Development*, https://doi.org/10.1002/ldr.3014.

Martinez, D. (2014) Traditional ecological knowledge, traditional resource management and silviculture in ecocultural restoration of temperate forests. In: M. Bozzano, R. Jalonen, E. Thomas, D. Boshier, L. Gallo, S. Cavers, S. Bordacs P. Smith, and J. Loo (eds), *Genetic Considerations in Ecosystem Restoration using Native Tree Species: State of the World's Forest Genetic Resources*, Food and Agriculture Organization of the United Nations and Biodiversity International, Rome, pp. 109–120.

Massey, C., Alpass, F., Flett, R., Lewis, K., Morriss, S. and Sligo, F. (2006) 'Crossing fields: the case of a multi-disciplinary research team', *Qualitative Research*, vol 6, no 2, pp. 131–147.

Mather, A.S. (1992) 'The forest transition', *Area*, vol 24, pp. 367–379.

McDonald, T., Gann, G.D., Jonson, J. and Dixon, K.W. (2016) *International Standards for the Practice of Ecological Restoration – Including Principles and Key Concepts*, Society for Ecological Restoration, Washington, DC.

McElwee, P. (2009) 'Reforesting "bare hills" in Vietnam: social and environmental consequences of the 5 million hectare reforestation program', *Ambio: A Journal of the Human Environment*, vol 38, no 6, pp. 325–333.

McNeill, D., Garcia Godos, J. and Gjerdåker, A. (2001) *Interdisciplinary Research on Development and the Environment*, Centre for Development and Environment, University of Oslo, Oslo.

Palmer, M.A., Falk, D.A. and Zedler, J.B. (2006) Ecological theory and restoration ecology. In: Falk, D.A., Palmer, M.A. and Zedler, J.B. (eds), *Foundations of Restoration Ecology*, Island Press, Washington, DC, pp. 3–26.

Pan, Y., Birdsey, R.A., Fang, J., Houghton, R., Kauppi, P.E., Kurz, W.A., Phillips, O.L., Shvidenko, A., Lewis, S.L., Canadell, J.G. and Ciais, P. (2011) 'A large and persistent carbon sink in the world's forests', *Science*, vol 333, pp. 988–993.

Pistorius, T. and Kiff, L. (2017) *From a biodiversity Perspective: Risks, Tradeoffs, and International Guidance for Forest Landscape Restoration*, UNIQUE, Freiburg.

Rudel, T.K., Coomes, O.T., Moran, E., Achard, F., Angelsen, A., Xu, J. and Lambin, E. (2005) 'Forest transitions: towards a global understanding of land use change', *Global Environmental Change*, vol 15, no 1, pp. 23–31.

Sabogal, C., Besacier, C. and McGuire, D. (2015) 'Forest and landscape restoration: concepts, approaches and challenges for implementation', *Unasylva*, vol 66, no 245, p. 3.

Sasaki, N. and Putz, F.E. (2009) 'Critical need for new definitions of forest and forest degradation in global climate change agreements', *Conservation Letters*, vol 2, pp. 226–232.

Stanturf, J., Lamb, D. and Madsen, P., eds. (2012a) *A Goal-Oriented Approach to Forest Landscape Restoration* (Vol. 16), Springer Science & Business Media, Dordrecht.

Stanturf, J., Lamb, D. and Madsen, P., eds. (2012b) *Forest Landscape Restoration: Integrating Natural and Social Sciences* (Vol. 15), Springer Science & Business Media, Dordrecht.

Steffen, W., Crutzen, P.J. and McNeill, J.R. (2007) 'The Anthropocene: are humans now overwhelming the great forces of nature', *AMBIO: A Journal of the Human Environment*, vol 36, no 8, pp. 614–621.

Thompson, I.D., Guariguata, M.R., Okabe, K., Bahamondez, C., Nasi, R., Heymell, V. and Sabogal, C. (2013) 'An operational framework for defining and monitoring forest degradation', *Ecology & Society*, vol 18, no 2, pp. 20–42.

UNESCO (undated) *Best Practices on Indigenous Knowledge BP.15*. Available at: www.unesco.org/most/bpik15.htm (accessed 10 March 2018).

Van den Besselaar, P. and Heimeriks, G. (2001) 'Disciplinary, multidisciplinary, interdisciplinary: concepts and indicators', paper presented at the 8th conference on Scientometrics and Informetrics – ISSI2001, Sydney, Australia, 16–20 July, pp. 705–716.

van Oosten, C., Uzamukunda, A. and Runhaar, H. (2018) 'Strategies for achieving environmental policy integration at the landscape level. A framework illustrated with an analysis of landscape governance in Rwanda', *Environmental Science & Policy*, vol 83, pp. 63–70.

Wortley, L., Hero, J.-M. and Howes, M. (2013) 'Evaluating ecological restoration success: a review of the literature', *Restoration Ecology*, vol 21, no 5, pp. 537–543.

WWF and IUCN (2000) *Minutes of the Forests Reborn Workshop in Segovia*, Unpublished.

2 Integration for restoration

Reflecting on lessons learned from the silos of the past

Rachel Carmenta and Bhaskar Vira

Introduction

The aim of this chapter is to discuss how a lack of integration across sectors, scales, disciplines and knowledges could inhibit progress towards effective forest landscape restoration (FLR). The discussion draws on evidence that elaborates on the shortfalls of approaching sustainable forest management through sectoral approaches. We focus in particular on the challenges that impact the management and governance of FLR interventions and those which confound and constrain the knowledge creation and the intellectual back-drop informing interventions and their implementation. These two issues are relevant to the fundamental questions that surround all forms of FLR: notably, FLR for what, pursued where and measured how (Mansourian *et al.*, 2017)?

We give special attention to the tropics as a site of rapid land use change involving forest loss and degradation and a geography targeted by the FLR agenda. Past experience in averting forest loss and degradation, and identifying the drivers associated with it, provides useful insights. We draw on our research experience and the literature with reference to examples from across the tropics and provide an illustrative case study from Indonesia's peatland management. In doing so, we aim to highlight the potential to inform the burgeoning interest in FLR interventions with experience and learning afforded by previous endeavours in the forest sector. Our analysis suggests that integration across sectors, spatial scales, knowledge systems and disciplines will be central to aligning and enabling the FLR agenda. We stress, in particular, the previous pot-holes on the road to forest restoration in order to highlight the need to incorporate these lessons in the current and emergent FLR agenda. We argue that without such reflection, the FLR momentum risks adopting an ahistorical and apolitical technocratic approach to the challenges of forest and land management. Solutions are emphasized in the latter sections of this volume.

Setting the stage: the implicit interdisciplinarity of forest landscape restoration

Momentum for FLR is building and visible in international and national policy arenas and the donor, UN, intergovernmental organization (IGO) and non-governmental organization (NGO) communities (Chapter 1). Indeed restoration can be seen on the ground and has already engaged numerous households and communities around the globe. The objective of FLR is immediately cross-sectoral and interdisciplinary, not least because forest landscapes encompass multiple land uses, interests and ecosystem services (Game *et al.*, 2014; Reed *et al.*, 2016; Bennett, 2017), but also because the clear aims of FLR specifically address social and ecological dimensions (Lamb, 2014). Landscapes are inherently dynamic, and therefore land use decisions have both intended and unintended consequences in other sectors and domains (Liu *et al.*, 2013; Phelps *et al.*, 2013; Veldman *et al.*, 2015). As such, attempts to manage and govern landscapes to new states of sustainability will demand an approach that can successfully engage such diversity (Reed *et al.*, 2016). In the next section, we give a brief overview of the limitations of silo-styled approaches to previous restoration and land use more generally.

Limitations of traditional restoration – the forester and the ecologist

Forest loss and degradation (as seen in Chapter 1) can frequently be traced back to static 'silo' thinking (failure to integrate across sectors, across disciplines and across spatial scales) manifest in institutions, management and governance arrangements. This conventional approach involves bottlenecks to integrated management that include disciplinary divides, sectoral approaches, entrenched interests, power relations and 'institutional stickiness', which have proven an insurmountable challenge to forest conservation and sustainability in many cases (Barr and Sayer, 2012; Sloan and Sayer, 2015).

The soaring popularity of FLR today can be understood as directly related to previous attempts to restore forests and avert forest loss. There are many examples spanning the tropics; they range from the fortress-style approach of protected areas to integrated conservation and development projects, community-based natural resource management, sustainable forest management, REDD+ (reducing emissions from deforestation and forest degradation, and the role of conservation, sustainable management of forests, and enhancement of forest carbon stocks in developing countries) and green growth (Agrawal and Gibson, 1999; Blom *et al.*, 2010; Dressler *et al.*, 2010; Barr and Sayer, 2012). Although numerous and widespread, these activities have not yet delivered the scale of success anticipated (Ferraro and Pattanayak, 2006). The 'restoration economy' and momentum surrounding the FLR agenda appears to face similar challenges, seeking also to reverse some of the same drivers of forest loss and degradation (Hosonuma *et al.*, 2012).

'Traditional restoration', under the purview of technicians, ecologists or foresters, has frequently taken the form of monoculture plantations or ecological restoration and is perhaps the most appropriate starting point to exemplify where a sectoral approach to restoration can fail. Focusing on either production or ecology may preclude identification (and subsequent management) of the initial drivers of species loss, and detract from managing the impacts of restoration across environmental and human dimensions (Lewis, 2001; Le *et al.*, 2014; Reed *et al.*, 2016; Bennett, 2017; Kodikara *et al.*, 2017). Traditional approaches to restoration are associated with a number of previously documented failures. These failures have included: plantations of exotic species associated with increased wildfire risk (Pausas and Keeley, 2014); establishment of tree species on grasslands with negative impacts on water provision (Veldman *et al.*, 2015); diseases affecting monocultures and killing large proportions of the plantation (Wingfield *et al.*, 2001); erroneously matching the biology of species to the biome and incurring mass failure (Lewis, 2001); and the introduction of invasive species at the expense of local ecosystems and associated livelihoods (Djoudi *et al.*, 2013).

Forest landscapes are increasingly expected to satisfy multiple goals and are sites where competing imperatives create the potential for trade-offs (Game *et al.*, 2013; Bennett and Chaplin-Kramer, 2016; Reed *et al.*, 2016). These can simultaneously include food production, energy, carbon sequestration, biodiversity conservation, livelihood provision, wellbeing, and infrastructure development and can involve commercial interests such as mining, industrial plantation expansion and estate crops (Laurance *et al.*, 2014; Sloan and Sayer, 2015; Franks *et al.*, 2017). Such complexity is typically not addressed in attempts to restore forests that are dominated by a single sector or disciplinary approach (Ostrom *et al.*, 2007). For example, improving conservation outcomes will likely require addressing food security and production, yet attempts to improve agricultural yields may have the most sustainable outcomes when coupled with mechanisms for forest conservation (Angelsen and Kaimowitz, 2001; Phalan *et al.*, 2016). Calls to expand beyond a dichotomous framing (i.e. of land sharing, land sparing – see Chapter 7, this volume) underline the potential to identify new solutions through reframing the debate with human wellbeing at the centre (Bennett, 2017). Managing restoration within the confines of a particular discipline or approach is unlikely to result in favourable outcomes, because drivers of species loss are often not confined to the biophysical system. Furthermore, social-ecological systems (SES) are not bounded, typically being interconnected and involving emergent properties in which the system is often more than the sum of its parts (Folke *et al.*, 2007 and see Chapter 5, this volume).

The interconnected nature of human–natural systems has been expressed in the ambition of the sustainable development goals (SDGs). SDG targets concerning both domains are clearly articulated, but are not holistically integrated and risk being treated in a piecemeal manner by agencies that continue to operate in isolated silos (Vira, 2015). The term FLR emerged in recognition that uncoordinated and non-integrated approaches have implications for the successful future management and governance of restoration within this broader framing of the SDGs, pushing towards multifunctional landscape management in which restoration is reconciled with other land uses. In this context, at least rhetorically, there is greater recognition of the need for integrated solutions.

Characterizing sectoral silos: drawing lessons from the past

Here we outline some of the key issues that have characterized 'silo'-styled approaches in the past, discuss how they detract from conservation outcomes, and caution that such lessons are valuable and necessary to transfer to the contemporary momentum towards FLR.

Interconnections and complexity influencing forest landscapes

Many of the pressures on forests come from decisions (and their ramifications) that are made beyond the forest sector (Geist and Lambin, 2002). Forest landscape restoration seeks to reverse long-standing trends of forest loss and degradation demanding adequate mitigation, or reversal, of the institutionalized incentives and preferences that drive deforestation (Geist and Lambin, 2002; Fischer *et al.*, 2012; Hosonuma *et al.*, 2012). Such drivers are often part of a larger system which is influenced by the political economic context, global markets and international agendas, operating at spatial scales that are typically beyond the locality of the site where landscape-level changes are sought (Geist and Lambin, 2002; Liu *et al.*, 2013). Further, complexity is exaggerated by scale-dependent phenomena, such as those related to trajectories of forest loss and land use change, and can be misdiagnosed if analysed on any one geospatial (Brondizio and Moran, 2012), temporal (Bhagwat *et al.*, 2014) or institutional scale alone (Sorrensen, 2003).

Land use decisions may involve deliberate trade-offs in which forests are committed to production due to an over-riding national priority of food security (Foley *et al.*, 2005; Franks *et al.*, 2017). Yet such ostensibly deliberate decisions often involve cascading effects in the form of environmental externalities (Foley *et al.*, 2005). Many of the externalities associated with land use and management have been largely invisible to conventional decision-making – the externalities associated with road expansion in the tropics is a classic example (e.g. deforestation, biodiversity loss, fire density) (Laurance *et al.*, 2014). Futhermore, feedback loops and interconnections between cause and effect are not always understood or predicted a priori.

Interventions may aim to mitigate a driver of forest loss and degradation (e.g. fire) and instead provoke forest loss through positive feedbacks (e.g. increased flammability through fuel build-up) (Pausas and Keeley, 2014) or redirect forest loss to other territories. International leakage challenges governance and monitoring arrangements to track multiple linkages in an increasingly connected world (Lambin *et al.*, 2018). The temporal breadth and geographic scope of remote sensing, particularly when linked to new technologies and big data, enable new possibilities to monitor progress transparently, identify transgressors, and create opportunities to improve the sustainability of commodity production. However, such systems alone are influenced by technological constraints, do not capture human dimensions and associated management decisions that constitute the patterns observed, and will not reorientate development pathways unless political support, institutional capacity and strong governance are also established (Liverman, 1998; Korhonen-Kurki *et al.*, 2014; Chazdon *et al.*, 2016; Gaveau *et al.*, 2017).

Policies to protect forests which are not coupled with policies to prevent displacement of deforestation can result in national and international leakage (Liu *et al.*, 2013). Additionally, policies to increase agricultural yields and spare land for nature face the risk of increasing land rents and potentially inducing run-away costs for conservation (Phelps *et al.*, 2013). Contradictory policies and perverse incentives have proven problematic for standing forests. These include agricultural and livestock subsidies or requiring forest land to be declared degraded before permits for plantations will be issued – thus inviting its degradation (Barr and Sayer, 2012).

Sectors divided

The significance of the political economic drivers of deforestation and degradation attest to their importance in reversing these processes in the pursuit of FLR. Yet, while progress has been made in quantifying the cost of degradation, and the externalities associated with deforestation and land degradation have been shown to be disproportionate (Costanza *et al.*, 1997; Dasgupta, 2008; Bennett and Chaplin-Kramer, 2016), there have been few examples of successes where diverse interests (e.g. agriculture, cattle, mining, forests, ecosystem services and non-monetary values) have been aligned (Reed *et al.*, 2016). Instead, sectors remain divided, demonstrating a consistent lack of coordination; notably, ministries of forestry and agriculture prescribe resource management in (dislocated) parallel domains of influence while issuing policy that pertains to an interconnected landscape and resource (Franks *et al.*, 2017).

Such dislocations are aggravated where ministries have overlapping authority and decentralized governance systems are in place (Anderson *et al.*, 2016). For example, in Indonesia, the Philippines and elsewhere, land allocation is a political and contested process in which stakeholders with vested interests and uneven power distribution have little incentive for integration or collaboration across sectors; rather, ministries compete for resources managed on short time frames, which hampers FLR efforts (Barr and Sayer, 2012; Baker *et al.*, 2014). Official land cover designations at the national level (e.g. forests) may manifest as distinct land uses at the local level (e.g. agriculture), yet reclassifications of the forest estate may be unattractive, in part due to the exchange of power and influence between levels of government that such reclassifications would entail (Chazdon *et al.*, 2016).

Some observers argue that there is now great potential to strengthen the strategies and secure the incentives for enabling FLR by subsuming FLR in climate mitigation strategies, in the sustainable development agenda, and in associated green growth plans (Ding *et al.*, 2017). Specifying the multiple benefits of FLR (e.g. job creation, food security, biodiversity conservation) may enable the blending of different sources of capital, such as climate, conservation and development finance (Ding *et al.*, 2017), yet coordinating such ambitions is essential if objectives are to be realistic and set within the geographic confines of the extent of national land estates (Franks *et al.*, 2017). There is a strong belief that what the forest sector has lacked is money, reflected in a considerable drive towards new forms of investment, and funding from carbon markets has shown marked success. Yet, there is very little evidence that money, in itself, will be sufficient if channelled through non-integrated approaches, as this does not address the institutional challenges that impede effective conservation or restoration activities. Moreover, pouring money into a sector that is sometimes plagued with challenges of corruption, insecure tenure and poor governance could result in perverse outcomes (Barr and Sayer, 2012).

Even if the major financial incentives necessary for a radical transformation of sectoral style management were to be operationalized, the complexity of social-ecological systems remains a significant challenge to resource managers. Divergent and competing interests in forest landscapes and interactions across geographic scales that manifest in a given area make it challenging for resource managers to adequately design governance systems which address complex interactions between cause, effect and the emergent properties of the system (Foley *et al.*, 2005; Folke *et al.*, 2007). Such system dynamics underscore contemporary approaches and concepts related to governance (e.g. polycentric and multi-level governance, integrated landscape approaches and ecosystem services). Integrated landscape approaches are gathering interest due to their potential to broker contested resource management decisions through a model that fosters cross-sectoral collaboration and encourages multi-stakeholder negotiation in which competing policies and interests can be reconciled; yet, without supporting legislation, finance or policy instruments, it remains unclear how such an approach can achieve radical institutional reform involving reshaping sectoral approaches at scale (Fischer *et al.*, 2012).

Definitions, criteria and the salience of semantics

While disciplinary differences are expected, concepts and terminology that appear straightforward at first glance can also be contested and complex. Semantics matter for forest landscape restoration. Definitions of key concepts such as forests, landscapes and restoration are not consistent even within contemporary 'high' levels of governance, and are similarly divergent between situated understandings and locally ascribed meanings (Dove, 1986; McElwee, 2009; Vasan, 2014; Chazdon *et al.*, 2016). Interpretations of these terms influence the methods to measure and inform FLR, yet they diverge depending on the viewpoint of policymakers, decision-makers, social scientists, natural scientists, foresters, farmers, land management organizations and NGOs (Dove, 1986; McElwee, 2009; Baker *et al.*, 2014; Chapter 9, this volume). These works show how definitions are politically relevant and can imbue power relations and further some political agendas while subjugating others. The notion of forest estate and degraded forests are political constructs, which serve to enable control over and appropriation of resources. In India, the term 'wasteland' was defined not in terms of its ecological productivity but rather, in terms of its ability to contribute revenues to the colonial state, while in Indonesia, marginal lands devoid of productivity and described as dominated by 'weeds' by the state were appropriated from local resource users (Dove, 1986; Baka, 2014).

The issue of definitions raises additional and important questions related to who decides what and where to restore, with what, for what benefit, to be shared by whom, and where (Veldman *et al.*, 2015; Mansourian *et al.*, 2017); each of which has important implications not only for integration across sectors and scales, but also for what forms of knowledge production, priorities and science are used to inform the development of programmes (Chapter 12, this volume). The criteria for targeting interventions within a landscape will not be sufficient if limited to ecological traits alone. Co-constructing definitions may be a part of building towards improved legitimacy, equity, relevance and social acceptance, which all contribute to the success of an initiative (Mansourian *et al.*, 2014). Yet, significant research questions remain, such as how the variety of experience can be generalized to indicators that are locally relevant and context sensitive, yet also applied at larger scales in other contexts (Sterling *et al.*, 2017).

Monitoring progress on FLR will demand interdisciplinary indicators and processes, since FLR covers both ecological restoration and human wellbeing beyond income metrics and over an extended time frame (Le *et al.*, 2012; Nguyen *et al.*, 2016). Arguments in development aid are relevant and highlight the importance of impact rather than input or output indicators – for example, simple accounts of hectares covered or seedlings distributed are not informative, potentially miss opportunities for incentivizing more long-standing commitments, and focus on quantity over quality (Mansourian *et al.*, 2017). Importantly, recent calls for locally solicited indicators that capture context-specific notions of wellbeing and a good life, and local perceptions of impact, may offer more meaningful metrics of impact at the local level (Sterling *et al.*, 2017).

Aligning with local realities

Local actors are crucial contributors to the performance of FLR interventions in most cases; where the engagement of different actors has not taken place, and the impacts of interventions on local actors have not been addressed, initiatives can fail (Nguyen *et al.*, 2016). Community dialogue and co-design during intervention development and implementation, as well as in evaluation (where important feedback can be incorporated into future iterations of the programme), are key to long-term FLR. International agendas, including FLR, entail projecting global objectives onto local-level realities through national and jurisdictional governance structures – a process that can involve reinterpretations at each of the different scales of governance (Vasan, 2014; Anderson *et al.*, 2016). True co-design represents a significant challenge in which distinct worldviews, priorities and aspirations come into contact and may require defining and negotiating new benchmarks of progress (Brown, 2005; Fischer *et al.*, 2012).

Further, the legacy of past interventions demonstrates the potential for externally designed interventions to be misaligned with local realities and generate significant and sometimes severe impacts for people and nature, raising questions related to environmental justice (Dressler *et al.*, 2010; Carmenta *et al.*, 2013; Chapter 4, this volume). Examples include where interventions have involved the displacement of local people, rights abuses, inequitable compensation and elite capture, and transformation of traditional land use practices, livelihoods, cultural norms and place-based attachments (Barr and Sayer, 2012; Djoudi *et al.*, 2013; Baker *et al.*, 2014). In these circumstances, new FLR initiatives have to overcome these past legacies of exclusionary, non-participatory and potentially exploitative practices, and find ways to build trust and to demonstrate that new approaches will genuinely break from these potentially entrenched ways of working.

Knowledge and value systems

Contemporary calls for interdisciplinary knowledge generation (e.g. sustainability science, conservation science and social-ecological systems) are relevant to the integrated remit of FLR (Foley *et al.*, 2005; Ostrom *et al.*, 2007; Bennett, 2017; Keeler *et al.*, 2017). Within such approaches, knowledge generation is less the sole responsibility of academics and scholars, foresters and ecologists, and instead engages multiple types of knowing – ranging from formal scientific traditions to more informal, place-based and experiential forms of knowledge, perceptions, worldviews and values held by diverse stakeholders and within traditional and indigenous knowledge systems and gender-sensitive approaches (Bennett *et al.*, 2017; Keeler *et al.*, 2017; Pascual *et al.*, 2017; Sterling *et al.*, 2017; Chapter 3, this volume). Biocultural approaches to conservation have attempted to reorientate conservation and development initiatives with a grounded framing in situated knowledges and value systems (Sterling *et al.*, 2017), thus averting the pitfalls of imposing understandings of cause, effect, reward and sanction and generating perverse and potentially unethical outcomes (e.g. motivational crowding out in Payment for Ecosystem Services (PES) schemes, i.e. incentives offered to farmers or landowners in exchange for managing their land to provide some sort of ecological service) (Muradian *et al.*, 2013). The Intergovernmental science-policy Platform on Biodiversity and Ecosystem Services (IPBES) framework offers an approach to capturing intrinsic, relational and place-based knowledge situated in pluralistic value systems attributed to forest landscapes (Pascual *et al.*, 2017). Such value systems may have no market value and have so far been obscured, or overlooked by conservation and development agendas (Hausmann *et al.*, 2016; Pascual *et al.*, 2017).

Local knowledge systems are generating increased recognition, and scholars from different orientations can collectively inform powerful, timely, policy-orientated solutions through fostering new partnerships and collaboration, including with those stakeholders beyond research traditions in transdisciplinary approaches to engagement, learning, outreach and policy change (Keeler *et al.*, 2017). Sustainability science orientated to real-world problems demands collaboration between disciplines, engaging communication across epistemologies, requiring reflexivity on the differentially endorsed methodological approaches to examining research problems across disciplinary conventions and navigating incongruent writing conventions and associated vocabularies (Sandbrook *et al.*, 2013; Haider *et al.*, 2018). Achieving an appropriate balance of 'methodological groundedness and epistemological agility' is challenging and calls for institutional innovation in the academy to incentivize sustainability science (Keeler *et al.*, 2017; Haider *et al.*, 2018).

Quantifying the costs and benefits of interventions transparently is necessary if deliberate choices are to be made regarding actions and trade-offs. However, modelling the financial implications of interventions, even across sectors, speaks to a particular worldview and does not deliver a cross-cultural conservation ethic (Fischer *et al.*, 2012; Wali *et al.*, 2017). Incorporating diverse value systems, potentially those outside of mainstream economic paradigms, in the design of interventions through meaningful participation remains a key challenge facing FLR (Game *et al.*,

2013). The rhetoric of a restoration economy in which employment and economic benefits will emerge through FLR (Ding *et al.*, 2017) suggests that the worldviews, aspirations and everyday practices of those to be impacted need to be considered, and where these are held in tension, equitably negotiated.

Superimposing value systems raises ethical considerations, and where perceptions of need and value clash, reforestation initiatives can fail. For example, in India in the 1970s, the perception of a fuelwood crisis led to the adoption of social forestry programmes designed to promote domestic fuelwood plantations. In practice, these were adopted at scale by farmers, who selected fast-growing, commercially valuable species (eucalyptus) instead of multifunctional species (Saxena, 1994). The failure to recognize the multiple uses of forest plantations, and their values to different stakeholders, had little impact on the availability of fuelwood; rather, it generated social protest (Casson, 2013).

Power, interests and diverse stakeholders

Integrated approaches improve the prospects for long-term permanence, yet reconciling all interests represents a key challenge for integrated landscape management and governance of coupled systems of humans and nature (Folke *et al.*, 2007; Game *et al.*, 2013). Landscape-scale interventions necessarily engage diverse stakeholders, who can differentially support and have access to powerful agenda (and national priorities). In this context, a key concern is reconciling alternative pathways for development (Vira, 2015; Carmenta *et al.*, 2017). While private sector engagement has been identified as a critical challenge to forest conservation, a full pursuit of the FLR agenda may require reframing debates about sustainability, engaging non-Western belief systems and values, and necessitate bold sustainability reforms, which potentially challenge existing interests in the status quo.

The buy-in of diverse actors is most likely when an intervention is perceived as legitimate to all. This is more likely where meaningful dialogue, facilitated by boundary partners and tools for assessing competing scenarios, has taken place and influenced the intervention (Game *et al.*, 2013). Civil society is important where nation states are weak, and expanded policy arenas that reflect new modalities of governance (as distinct from governments) may enable a more inclusive and participatory approach, which draws in these diverse stakeholders (Fischer *et al.*, 2012).

Indonesian peatland restoration: an illustrative case study

Indonesia's carbon-dense tropical peat swamp forests represent a globally significant contemporary hotspot of land use change (Turetsky *et al.*, 2015). Previously unexploited, such peatlands are now contested, fire-prone frontiers undergoing fast-paced and drastic human-induced transformation to oil palm agriculture (Koh *et al.*, 2011). The landscape transformation is associated with the arrival of new technologies (e.g. drainage canals, heavy machinery and industrial plants), land user groups (e.g. agro-industry, mid-sector investment, migrant and transmigrant communities) and bureaucrats charged with management (e.g. provincial and district officials). In Riau province, Sumatra, the extensive land use change to lucrative oil palm and plantation crop cultivation is a process that necessitates drainage, which is associated with wildfires (Figure 2.1). Peatland fires generate disproportionate burdens across sectors and scales (Marlier *et al.*, 2013; Turetsky *et al.*, 2015; Koplitz *et al.*, 2016; World Bank, 2016; Wijedasa *et al.*, 2017). Large-scale conversion has included the ill-fated agricultural 'Mega-Rice' project, which drained 1 million ha of peatland for rice production, produced not a single grain, and has burned extensively and repeatedly since 1997 (Hoscilo *et al.*, 2011). In previously forested and flooded peatlands, fires are now annual events, decoupled from climate anomalies (Gaveau *et al.*, 2014).

The magnitude of the fires has spurred a range of fire management interventions, which have suffered chronic implementation failures, are highly contested between stakeholder groups and have been the source of diplomatic tensions throughout the Association of Southeast Asian Nations (ASEAN) region (Tacconi, 2016; Carmenta *et al.*, 2017). A new range of responses to the 2015 mega-fire events have mobilized considerable interest

Figure 2.1 Indonesian peatland fires smouldering in Kalimantan. The peatland fires of 2015 caused inordinate burdens across sectors and scales and catalysed a new wave of responses to improved peatland management, including commitments to restore 2 million hectares of peat.

Source: photo © Björn Vaughn.

from diverse communities (e.g. government, NGO, private and donor) to implement restoration activities, including interventions supporting the introduction of alternative paludiculture (water-tolerant farming systems) that do not require fire, the restoration of peatlands through the Global Peatlands Initiative (GPI), and the restoration and rewetting of 2 million hectares of peatland under the auspices of the newly created Peatland Restoration Agency (BRG, Indonesian acronym) (Figure 2.2). These contemporary and previous restoration activities on degraded Indonesian peatland illustrate some of the challenges associated with silo-style approaches in complex landscapes, and are expanded on in the following.

The restoration attempts on Indonesia's peatlands offer an example of where biology and species physiology are key to getting restoration right – only water-tolerant species will persist, and full restoration requires the peatland substrate itself to be cool enough for the soil microbial community to re-establish and thrive following reintroduction via inoculation. Yet, even with these key factors in place, the hydrology of peatlands cannot be ignored; it requires additional expertise for adequate management. While the natural sciences are important, external stressors to forest conservation and restoration interventions such as climate change, fire and market forces can jeopardize even the most biologically accurate interventions (Barlow *et al.*, 2012; DeFriess *et al.*, 2015; Kodikara *et al.*, 2017), and institutional mechanisms to deal with such forces demand inclusion.

Any approach to restoration must seek to identify and engage the stakeholders and sectors whose actions, policies and management strategies influence the landscape of interest. Peatland restoration in Indonesia has faced significant start-up challenges, in part due to the lack of integration between sectors, scales and stakeholders. For example, in the wake of 2015, the Ministry of Public Works and Housing continued to plan for peatland drainage, while the Ministry of Environment and Forestry had restricted such practices. Following the 2015 fire event, the Ministry of Public Works has exercised its authority to close some of the major drainage canals, yet smaller canals are under the jurisdiction of provincial and local administrations. Successful approaches to restoration require engagement of ministries and integration between levels of governance structures in the context of adequate enforcement. Such endeavours necessitate aligned definitions (e.g. of restoration, rewetting and deep peat). Indonesia has notoriously lacked both an agreed map of its peatland estate and a unified map of land tenure, both of which are currently commissioned (i.e. the Peat Prize and 'One Map' initiatives) (Barr and Sayer, 2012; Alisjahbana and Busch, 2017).

Ministries coordinating activities both horizontally and vertically, orientated around clear definitions and systematic data sources, supported by sound biology and strong governance, is likely not enough. Attempts to rewet and restore degraded Indonesian peatlands through canal blocking affect the livelihoods of local people and the assets of large plantation

Figure 2.2 Canal blocking in Riau, Sumatra. The large fires in 2015 have restored interest in fire mitigation, a large part of which is being pursued through canal blocking as an attempt to rewet and restore degraded and fire-prone peatlands.

Source: photo © Rachel Carmenta.

companies whose crops do not tolerate water-logged conditions (Figure 2.2). Canal-blocking activities have been sabotaged by local people defending their right to navigate drainage canals. The BRG can only achieve its aims with the buy-in, interest and trust of these stakeholders. Similarly, the BRG requires companies to restore parts of their estates and has been met with intense lobbying from powerful private sector stakeholders (Alisjahbana and Busch, 2017). The BRG has no legal force but is supported by controversial presidential instructions (i.e. ministerial decree No. 17/2017), which demand the rewetting (as part of the restoration process) of cleared but unplanted concession areas, and a restriction of agricultural activities to the current harvest cycle (i.e. after harvest the peatland must be restored). The case is being lobbied, and commentators speculate over how newcomers (and fire) will be kept out, if companies are forced to withdraw from peatlands.

New tools supporting integrated approaches to restoration are being developed and could prove promising for peatland restoration. Such tools include interactive models that make various trade-offs between different land use scenarios explicit and have engaged multiple stakeholders and disciplinary expertise to ensure their breadth and increase the chance of uptake by key decision-makers. For example, a contemporary holistic assessment of the extended burdens (e.g. those related to sectors of health, education, environment, etc.) of the peatland fires is being modelled by an interdisciplinary team of scientists engaging with diverse experts to inform a scenario decision-making tool. The tool enables peatland management scenarios, and their implicit trade-offs, to be transparently addressed by stakeholders in the policy arena (Marlier *et al.*, personal communication). The public benefits of restoration may benefit from being approached in a similar way (Ding *et al.*, 2017).

Conclusions: landscape of possibility hinges on learning from the past

The restoration agenda heralds significant, albeit with many caveats, hope for sustainable futures, and holds the potential to make a positive contribution to local livelihoods, wellbeing and the ecological integrity of forest landscapes while contributing to securing international and national policy objectives (e.g. SDGs, the Convention on Biodiversity's Aichi targets and food security). Yet FLR is active amid multiple and competing demands for land and resources, which are too often acting in parallel rather than in unison. Silo-style approaches in which ecologists produce narrow restoration guidelines, foresters plant trees for production, companies invest in tree planting for carbon credits, and so on must be exchanged for new models of governance, management and knowledge production.

The long-term, large-scale and multiple objectives of FLR dictate that a silo approach will not work; rather, institutional reform and transformational change are paramount. An integrated approach is necessary for all aspects of an FLR agenda – for design, knowledge production, governance, management and monitoring. However, experience shows the potential to underperform, and there is a need to learn from the past to inform the future, particularly at a time when a multitude of new actors are engaging in the FLR agenda. The present landscape of possibility will require that significant challenges are overcome, for example institutional stickiness, vested interests, inadequate incentives and static sectoral designs, which have hampered forest conservation in the past.

The conventional focus on ecological restoration and associated lack of coordination and buy-in across sectors, lack of integration across scales of governance and knowledges, and failure to engage with the political economic context within which forest and land use decisions are made have hampered progress to date (Barr and Sayer, 2012; Mansourian *et al.*, 2014; Sloan and Sayer, 2015; Anderson *et al.*, 2016; Bennett and Chaplin-Kramer, 2016). A more successful model might be capable not only of integrating scales but also of mediating and satisfying the competing interests for land and land use sustainably into the future (Fischer *et al.*, 2012; Bennett and Chaplin-Kramer, 2016; Bennett, 2017). FLR is unlikely to serve as a panacea alone and will need to engage with additional strategies, including those in consuming countries that address social norms, consumption patterns and diet preferences, if a sustainable future is to be secured for all (Fischer *et al.*, 2012).

The need for integration across sectors and scales is increasingly recognized, perhaps now more than ever in our telecoupled world (Liu *et al.*, 2013). Rising global population and associated commodity demand increase the human footprint on the planet in an increasingly neoliberal economy (Foley *et al.*, 2005; Dasgupta, 2008; Raworth, 2017), resulting in concern about the significance of human-induced global environmental change in the context of irreversible tipping points facing planetary systems (Rockström *et al.*, 2009; Steffen *et al.*, 2015). The FLR agenda comes with transformational potential, which can reverse some of these trends, but its effectiveness will be highly constrained unless it recognizes the lessons of the past, most notably the need for integrated and holistic ways of promoting changes within interlinked social and ecological systems.

References

Agrawal, A. and Gibson, C. C. (1999) 'Enchantment and disenchantment: The role of community in natural resource conservation', *World Development*, vol 27, no 4, pp. 629–649.

Alisjahbana, A. S. and Busch, J. M. (2017) 'Forestry, forest fires, and climate change in Indonesia', *Bulletin of Indonesian Economic Studies*, vol 53, no 2, pp. 111–136.

Anderson, Z. R., Kusters, K., McCarthy, J. and Obidzinski, K. (2016) 'Green growth rhetoric versus reality: Insights from Indonesia', *Global Environmental Change*, vol 38, pp. 30–40.

Angelsen, A. and Kaimowitz, D. (2001) *Agricultural technologies and tropical deforestation*. Oxford, UK: CABI.

Baka, J. (2014) 'What wastelands? A critique of biofuel policy discourse in South India', *Geoforum*, vol 54, pp. 315–323.

Baker, S., Eckerberg, K. and Zachrisson, A. (2014) 'Political science and ecological restoration', *Environmental Politics*, vol 23, no 3, pp. 509–524.

Barlow, J., Parry, L., Gardner, T. A., Ferreira, J., Aragão, L. E., Carmenta, R., Berenguer, E., Vieira, I. C., Souza, C. and Cochrane, M. A. (2012) 'The critical importance of considering fire in REDD+ programs', *Biological Conservation*, vol 154, pp. 1–8.

Barr, C. M. and Sayer, J. A. (2012) 'The political economy of reforestation and forest restoration in Asia-Pacific: Critical issues for REDD+', *Biological Conservation*. vol 154, pp. 9–19.

Bennett, E. M. (2017) 'Changing the agriculture and environment conversation', *Nature Ecology and Evolution*, vol 1, no 1, pp. 1–2.

Bennett, E. M. and Chaplin-Kramer, R. (2016) 'Science for the sustainable use of ecosystem services', *F1000Research*, vol 5, p. 2622.

Bennett, N. J., Roth, R., Klain, S. C., Chan, K., Christie, P., Clark, D. A., Cullman, G., Curran, D., Durbin, T.J., Epstein, G. and Greenberg, A. (2017) 'Conservation social science: Understanding and integrating human dimensions to improve conservation', *Biological Conservation*, vol 205, pp. 93–108.

Bhagwat, S. A., Nogué, S. and Willis, K. J. (2014) 'Cultural drivers of reforestation in tropical forest groves of the Western Ghats of India', *Forest Ecology and Management*, vol 329, pp. 393–400.

Blom, B., Sunderland, T. and Murdiyarso, D. (2010) 'Getting REDD to work locally: Lessons learned from integrated conservation and development projects', *Environmental Science and Policy*, vol 13, no 2, pp. 164–172.

Brondizio, E. S. and Moran, E. F. (2012) 'Level-dependent deforestation trajectories in the Brazilian Amazon from 1970 to 2001', *Population and Environment*, vol 34, no 1, pp. 69–85.

Brown, K. (2005) 'Addressing trade-offs in forest landscape restoration', in Mansourian, S., Dudley, N. and Vallauri, D., eds. *Forest restoration in landscapes: Beyond planting trees*. New York: Springer, pp. 59–62.

Carmenta, R., Vermeylen, S., Parry, L. and Barlow, J. (2013) 'Shifting cultivation and fire policy: Insights from the Brazilian Amazon', *Human Ecology*, vol 41, no 4, pp. 603–614.

Carmenta, R., Zabala, A., Daeli, W. and Phelps, J. (2017) 'Perceptions across scales of governance and the Indonesian peatland fires', *Global Environmental Change*, vol 46, pp. 50–59.

Casson, A. (2013) *The controversy surrounding eucalypts in social forestry programs of Asia*, working paper, National Centre for Development Studies – The Australian National University, Canberra.

Chazdon, R. L., Brancalion, P. H., Laestadius, L., Bennett-Curry, A., Buckingham, K., Kumar, C., Moll-Rocek, J., Vieira, I. C. G. and Wilson, S. J. (2016) 'When is a forest a forest? Forest concepts and definitions in the era of forest and landscape restoration', *Ambio*, vol 45, no 5, pp. 538–550.

Costanza, R., d'Arge, R., De Groot, R., Farber, S., Grasso, M., Hannon, B., Limburg, K., Naeem, S., O'Neill, R. V., Paruelo, J. and Raskin, R. G. (1997)

'The value of the world's ecosystem services and natural capital', *Nature*, vol 387, no 6630, p. 253.

Dasgupta, P. (2008) 'Nature in economics', *Environment and Resource Economics*, vol 39, no 1, pp. 1–7.

DeFries, R., Rudel, T. K., Uriarte, M. and Hansen, M. (2010) 'Deforestation driven by urban population growth and agricultural trade in the twenty-first century', *Nature Geoscience*, vol 3, pp. 178–181.

Ding, H., Faruqi, S., Wu, A., Altamirano, J. C., Anchondo Ortega, A., Verdone, M. and Vergara, W. (2017) *Roots of prosperity: The economics and finance of restoring land*. Washington, DC: WRI.

Djoudi, H., Brockhaus, M. and Locatelli, B. (2013) 'Once there was a lake: Vulnerability to environmental changes in northern Mali', *Regional Environmental Change*, vol 13, no 3, pp. 493–508.

Dove, M. (1986) 'The practical reason of weeds in Indonesia: Peasant vs. state views of Imperata and Chromolaena', *Human Ecology*, vol 14, pp. 163–190.

Dressler, W., Büscher, B., Schoon, M., Brockington, D. A. N., Hayes, T., Kull, C. A., McCarthy, J. and Shrestha, K. (2010) 'From hope to crisis and back again? A critical history of the global CBNRM narrative', *Environmental Conservation*, vol 37, no 1, pp. 5–15.

Ferraro, P. J. and Pattanayak, S. K. (2006) 'Money for nothing? A call for empirical evaluation of biodiversity conservation investments', *PLoS Biology*, vol 4, no 4, pp. 482–488.

Fischer, J., Dyball, R., Fazey, I., Gross, C., Dovers, S., Ehrlich, P. R., Brulle, R. J., Christensen, C. and Borden, R. J. (2012) 'Human behavior and sustainability', *Frontiers in Ecology and the Environment*, vol 10, no 3, pp. 153–160.

Foley, J. A., DeFries, R., Asner, G. P., Barford, C., Bonan, G., Carpenter, S. R., Chapin, F. S., Coe, M. T., Daily, G. C., Gibbs, H. K. and Helkowski, J. H. (2005) 'Global consequences of land use', *Science*, vol 309, no 5734, pp. 570–574.

Folke, C., Pritchard Jr, L., Berkes, F., Colding, J. and Svedin, U. (2007) 'The problem of fit between ecosystems and institutions: Ten years later', *Ecology and Society*, vol 12, no 1, article 30. Available at: www.ecologyandsociety.org/vol12/iss1/art30/ (accessed 19 June 2018).

Franks, P., Hou-Jones, X., Fikreyesus, D., Sintayehu, M., Mamuye, S., Danso, E. Y., Meshack, C. K., McNicol, I. and Soesbergen, A. V. (2017) *Reconciling forest conservation with food production in sub-Saharan Africa: Case studies from Ethiopia, Ghana and Tanzania*. London: IIED.

Game, E. T., Meijaard, E., Sheil, D. and McDonald-Madden, E. (2014) 'Conservation in a wicked complex world; challenges and solutions', *Conservation Letters*, vol 7, no 3, pp. 271–277.

Gaveau, D. L., Salim, M. A., Hergoualc'h, K., Locatelli, B., Sloan, S., Wooster, M., Marlier, M. E., Molidena, E., Yaen, H., DeFries, R. and Verchot, L. (2014) 'Major atmospheric emissions from peat fires in Southeast Asia during non-drought years: Evidence from the 2013 Sumatran fires', *Scientific Reports*, vol 4, p. 6112.

Gaveau, D. L., Pirard, R., Salim, M. A., Tonoto, P., Yaen, H., Parks, S. A. and Carmenta, R. (2017) 'Overlapping land claims limit the use of satellites to monitor no-deforestation commitments and no-burning compliance', *Conservation Letters*, vol 10, no 2, pp. 257–264.

Geist, H. J. and Lambin, E. F. (2002) 'Proximate causes and underlying driving forces of tropical deforestation', *BioScience*, vol 52, no 2, p. 143.

Haider, L. J., Hentati-Sundberg, J., Giusti, M., Goodness, J., Hamann, M., Masterson, V. A., Meacham, M., Merrie, A., Ospina, D., Schill, C. and Sinare, H. (2018) 'The undisciplinary journey: Early-career perspectives in sustainability science', *Sustainability Science*, vol 13, no 1, pp. 191–204.

Hausmann, A., Slotow, R. O. B., Burns, J. K. and Di Minin, E. (2016) 'The ecosystem service of sense of place: Benefits for human well-being and biodiversity conservation', *Environmental Conservation*, vol 43, no 2, pp. 117–127.

Hoscilo, A., Page, S. E., Tansey, K. J. and Rieley, J. O. (2011) 'Effect of repeated fires on land-cover change on peatland in southern Central Kalimantan, Indonesia, from 1973 to 2005', *International Journal of Wildland Fire*, vol 20, no 4, pp. 578–588.

Hosonuma, N., Herold, M., De Sy, V., De Fries, R. S., Brockhaus, M., Verchot, L., Angelsen, A. and Romijn, E. (2012) 'An assessment of deforestation and forest degradation drivers in developing countries', *Environmental Research Letters*, vol 7, no 4, pp. 044009.

Keeler, B. L., Chaplin-Kramer, R., Guerry, A. D., Addison, P. F., Bettigole, C., Burke, I. C., Gentry, B., Chambliss, L., Young, C., Travis, A. J. and Darimont, C. T. (2017) 'Society is ready for a new kind of science – Is academia?', *BioScience*, vol 67, no 7, pp. 591–592.

Kodikara, K. A. S., Mukherjee, N., Jayatissa, L. P., Dahdouh-Guebas, F. and Koedam, N. (2017) 'Have mangrove restoration projects worked? An in-depth study in Sri Lanka', *Restoration Ecology*, vol 25, no 5, pp. 705–716.

Koh, L. P., Miettinen, J., Liew, S. C. and Ghazoul, J. (2011) 'Remotely sensed evidence of tropical peatland conversion to oil palm', *Proceedings of the National Academy of Sciences*, vol 108, no 12, pp. 5127–5132.

Koplitz, S. N., Mickley, L. J., Marlier, M. E., Buonocore, J. J., Kim, P. S., Liu, T., Sulprizio, M. P., DeFries, R. S., Jacob, D. J., Schwartz, J. and Pongsiri, M. (2016) 'Public health impacts of the severe haze in Equatorial Asia in September–October 2015: Demonstration of a new framework for informing fire management strategies to reduce downwind smoke exposure', *Environmental Research Letters*, vol 11, no 9, pp. 094023.

Korhonen-Kurki, K., Sehring, J., Brockhaus, M. and Di Gregorio, M. (2014) 'Enabling factors for establishing REDD+ in a context of weak governance', *Climate Policy*, vol. 14, no 2, pp. 167–186.

Lamb, D. (2014) *Large-scale forest restoration*. London: Earthscan.

Lambin, E. F., Gibbs, H. K., Heilmayr, R., Carlson, K. M., Fleck, L. C., Garrett, R. D., de Waroux, Y. L. P., McDermott, C. L., McLaughlin, D., Newton, P. and Nolte, C. (2018) 'The role of supply-chain initiatives in reducing deforestation', *Nature Climate Change*, vol. 8, pp. 109–116.

Laurance, W. F., Sayer, J. and Cassman, K. G. (2014) 'Agricultural expansion and its impacts on tropical nature', *Trends in Ecology and Evolution*, vol 29, no 2, pp. 107–116.

Le, H. D., Smith, C., Herbohn, J. and Harrison, S. (2012) 'More than just trees: Assessing reforestation success in tropical developing countries', *Journal of Rural Studies*, vol 28, no 1, pp. 5–19.

Lewis III, R. R. (2001) 'Mangrove restoration – Costs and benefits of successful ecological restoration', *Proceedings of the Mangrove Valuation Workshop,*

Universiti Sains Malaysia, Penang, 4–8 April, 2001. Beijer International Institute of Ecological Economics, Stockholm, Sweden.

Liu, J., Hull, V., Batistella, M., DeFries, R., Dietz, T., Fu, F., Hertel, T., Izaurralde, R. C., Lambin, E., Li, S. and Martinelli, L. (2013) 'Framing sustainability in a telecoupled world', *Ecology and Society*, vol 18, no 2.

Liverman, D. (1998) *People and pixels: Linking remote sensing and social science.* Washington, DC: The National Academies Press.

Mansourian, S., Aquino, L., Erdmann, T. K. and Pereira, F. (2014) 'A comparison of governance challenges in forest restoration in Paraguay's privately-owned forests and Madagascar's co-managed state forests', *Forests*, vol 5, no 4, pp. 763–783.

Mansourian, S., Stanturf, J., Derkyi, M. and Engel, V. (2017) 'Forest landscape restoration: Increasing the positive impacts of forest restoration or simply the area under tree cover?', *Restoration Ecology*, vol 25, no 2, pp. 178–183.

Marlier, M. E., DeFries, R. S., Voulgarakis, A., Kinney, P. L., Randerson, J. T., Shindell, D. T., Chen, Y. and Faluvegi, G. (2013) 'El Niño and health risks from landscape fire emissions in southeast Asia', *Nature Climate Change*, vol 3, no 2, p. 131.

McElwee, P. (2009) 'Reforesting "bare hills" in Vietnam: Social and environmental consequences of the 5 million hectare reforestation program', *AMBIO: A Journal of the Human Environment*, vol 38, no 6, pp. 325–333.

Muradian, R., Arsel, M., Pellegrini, L., Adaman, F., Aguilar, B., Agarwal, B., Corbera, E., Ezzine de Blas, D., Farley, J., Froger, G., Garcia-Frapolli, E., Gómez-Baggethun, E., Gowdy, J. Kosoy, N., Le Coq, J. F., Leroy, P., May, P., Méral, P., Mibielli, P., Norgaard, R., Ozkaynak, B., Pascual, U., Pengue, W., Perez, M., Pesche, D., Pirard, R., Ramos-Martin, J., Rival, L., Saenz, F., Van Hecken, G., Vatn, A., Vira, B. and Urama, K. (2013) 'Payments for ecosystem services and the fatal attraction of win-win solutions', *Conservation Letters*, vol 6, no 4, pp. 274–279.

Nguyen, T. P., Nguyen Van Tam, Le Phat Quoi and Parnell, K. E. (2016) 'Community perspectives on an internationally funded mangrove restoration project: Kien Giang province, Vietnam', *Ocean and Coastal Management*, vol 119, pp. 146–154.

Ostrom, E., Janssen, M. A. and Anderies, J. M. (2007) 'Going beyond panaceas', *Proceedings of the National Academy of Sciences*, vol 104, no 39, pp. 15176–15178.

Pascual, U., Balvanera, P., Díaz, S., Pataki, G., Roth, E., Stenseke, M., Watson, R. T., Dessane, E. B., Islar, M., Kelemen, E. and Maris, V. (2017) 'Valuing nature's contributions to people: The IPBES approach', *Current Opinion in Environmental Sustainability*, vol 26, pp. 7–16.

Pausas, J. G. and Keeley, J. E. (2014) 'Abrupt climate-independent fire regime changes', *Ecosystems*, vol 17, no 6, pp 109–1120.

Phalan, B., Green, R. E., Dicks, L. V., Dotta, G., Feniuk, C., Lamb, A., Strassburg, B. B., Williams, D. R., Zu Ermgassen, E. K. and Balmford, A. (2016) 'How can higher-yield farming help to spare nature?', *Science*, vol 351, no 6272, pp. 450–451.

Phelps, J., Carrasco, L. R., Webb, E. L., Koh, L. P. and Pascual, U. (2013) 'Agricultural intensification escalates future conservation costs', *Proceedings of the National Academy of Sciences*, vol 110, no 19, pp. 7601–7606.

Raworth, K. (2017) 'A doughnut for the Anthropocene: Humanity's compass in the 21st century', *The Lancet Planetary Health*, vol 1, no 2, pp. e48–e49.

Reed, J., Van Vianen, J., Deakin, E. L., Barlow, J. and Sunderland, T. (2016) 'Integrated landscape approaches to managing social and environmental issues in the tropics: Learning from the past to guide the future', *Global Change Biology*, vol 22, no 7, pp. 2540–2554.

Rockström, J., Steffen, W., Noone, K., Persson, Å., Chapin III, F. S., Lambin, E., Lenton, T. M., Scheffer, M., Folke, C., Schellnhuber, H. J., Nykvist, B., de Wit, C. A., Hughes, T., van der Leeuw, S., Rodhe, H., Sörlin, S., Snyder, P. K., Costanza, R., Svedin, U., Falkenmark, M., Karlberg, L., Corell, R. W., Fabry, V. J., Hansen, J., Walker, B., Liverman, D., Richardson, K., Crutzen, P. and Foley, J. (2009) 'Planetary boundaries: Exploring the safe operating space for humanity', *Ecology and Society*, vol 14, no 2, p. 32.

Sandbrook, C., Adams, W. M., Büscher, B. and Vira, B. (2013) 'Social research and biodiversity conservation', *Conservation Biology*, vol 27, no 6, pp. 1487–1490.

Saxena, N. C. (1994) *India's eucalyptus craze: The god that failed.* New Delhi, India: SAGE Publications.

Sloan, S. and Sayer, J. A. (2015) 'Forest resources assessment of 2015 shows positive global trends but forest loss and degradation persist in poor tropical countries', *Forest Ecology and Management*, vol 352, no 7, pp. 134–145.

Sorrensen, C. L. (2003) 'Frontier spaces of vulnerability : Regional change, urbanization, drought and fire hazard in Santarem, Para, Brazil', *Environment*, vol 6, no 1, pp. 123–144.

Steffen, W., Richardson, K., Rockström, J., Cornell, S. E., Fetzer, I., Bennett, E. M., Biggs, R., Carpenter, S. R., De Vries, W., de Wit, C. A. and Folke, C. (2015) 'Planetary boundaries: Guiding human development on a changing planet', *Science*, vol 347, no 6223, pp. 1259855.

Sterling, E. J., Filardi, C., Toomey, A., Sigouin, A., Betley, E., Gazit, N., Newell, J., Albert, S., Alvira, D., Bergamini, N. and Blair, M. (2017) 'Biocultural approaches to well-being and sustainability indicators across scales', *Nature Ecology & Evolution*, vol 1, no 12, p. 1798.

Tacconi, L. (2016) 'Preventing fires and haze in Southeast Asia', *Nature Climate Change*, vol 6, pp. 640–643.

Turetsky, M. R., Benscoter, B., Page, S., Rein, G., van der Werf, G. R. and Watts, A. (2015) 'Global vulnerability of peatlands to fire and carbon loss', *Nature Geoscience*, vol 8, pp. 11–14.

Vasan, S. (2014) 'Ethnography of the forest guard: Contrasting discourses, conflicting roles and policy implementation', *Economic and Political Weekly*, vol 37, no 40, pp. 4125–4133.

Veldman, J. W., Overbeck, G. E., Negreiros, D., Mahy, G., Le Stradic, S., Fernandes, G. W., Durigan, G., Buisson, E., Putz, F. E. and Bond, W. J. (2015) 'Tyranny of trees in grassy biomes', *Science*, vol 347, no 6221, pp. 484–485.

Vira, B. (2015) 'Taking natural limits seriously: Implications for development studies and the environment', *Development and Change*, vol 46, no 4, pp. 762–776.

Wali, A., Alvira, D., Tallman, P., Ravikumar, A. and Macedo, M. (2017) 'A new approach to conservation: Using community empowerment for sustainable well-being', *Ecology and Society*, vol 22, no 4.

Wijedasa, L. S., Jauhiainen, J., Könönen, M., Lampela, M., Vasander, H., Leblanc, M. C., Evers, S., Smith, T. E. L., Yule, C. M., Varkkey, H., Lupascu, M., Parish, F., Singleton, I., Clements, G. R., Abdul Aziz, S., Harrison, M. E., Cheyne, S., Anshari, G. Z., Meijaard, E., Goldstein, J. E., Waldron, S., Hergoualc'h, K., Dommain, R., Frolking, S., Evans, C. D., Posa, M. R. C., Glaser, P. H., Suryadiputra, N., Lubis, R., Santika, T., Padfield, R., Kurnianto, S., Hadisiswoyo, P., Lim, T. W., Page, S. E., Gauci, V., Van Der Meer, P. J., Buckland, H., Garnier, F., Samuel, M. K., Choo, L. N. L. K., O'Reilly, P., Warren, M., Suksuwan, S., Sumarga, E., Jain, A., Laurance, W. F., Couwenberg, J., Joosten, H., Vernimmen, R., Hooijer, A., Malins, C., Cochrane, M. A., Perumal, B., Siegert, F., Peh, K. S.-H., Comeau, L.-P., Verchot, L., Harvey, C. F., Cobb, A., Jaafar, Z., Wösten, H., Manuri, S., Müller, M., Giesen, W., Phelps, J., Yong, D. L., Silvius, M., Wedeux, B. M. M., Hoyt, A., Osaki, M., Hirano, T., Takahashi, H., Kohyama, T. S., Haraguchi, A., Nugroho, N. P., Coomes, D. A., Quoi, L. P., Dohong, A., Gunawan, H., Gaveau, D. L. A., Langner, A., Lim, F. K. S., Edwards, D. P., Giam, X., Van Der Werf, G., Carmenta, R., Verwer, C. C., Gibson, L., Gandois, L., Graham, L. L. B., Regalino, J., Wich, S. A., Rieley, J., Kettridge, N., Brown, C., Pirard, R., Moore, S., Ripoll Capilla, B., Ballhorn, U., Chew Ho, C. H., Hoscilo, A., Lohberger, S., Evans, T. A., Yulianti, N., Blackham, G., Onrizal, Husson, S., Murdiyarso, D., Pangala, S., Cole, L. E. S., Tacconi, L., Segah, H., Tonoto, P., Lee, J. S. H., Schmilewski, G., Wulffraat, S., Indra Putra, E. I., Cattau, M. E., Clymo, R. S., Morrison, R., Mujahid, A., Miettinen, J., Liew, S. C., Valpola, S., Wilson, D., D'Arcy, L., Gerding, M., Sundari, S., Thornton, S. A., Kalisz, B., Chapman, S. J., Su, A. S. M., Basuki, I., Itoh, M., Traeholt, C., Sloan, S., Sayok, A. K. and Andersen, R. (2017) 'Denial of long-term issues with agriculture on tropical peatlands will have devastating consequences', *Global Change Biology*, vol 23, no 3, pp. 977–982.

Wingfield, M. J., Slippers, B., Roux, J. and Wingfield, B. D. (2001) 'Worldwide movement of exotic forest fungi, especially in the tropics and the southern hemisphere: This article examines the impact of fungal pathogens introduced in plantation forestry', *AIBS Bulletin*, vol 51, no 2, pp. 134–140.

World Bank. (2016) *The cost of fire: An economic analysis of Indonesia's 2015 fire crisis*. Jakarta, Indonesia.

3 Considering diverse knowledge systems in forest landscape restoration

Frank K. Lake, Christian P. Giardina,
John Parrotta and Iain Davidson-Hunt

Introduction

If forest landscape restoration (FLR) aims towards living sustainably within landscapes and restoring degraded socio-ecological systems, then integrating lessons of Traditional and Western knowledge systems can inform this effort (c.f. Ruiz-Mallén and Corbera, 2013). Knowledge systems represent much more than repositories of timeless information useful to today's managers and restorationists: they are intricately coupled human and natural systems that have evolved through intergenerational and community-based stewardship of natural resources. In this context, Berkes (2007) cautions that viewing community-based conservation as a panacea ignores the complexity and depth that must be considered when engaging communities and their knowledge in conservation activities. Rather, effective and equitable strategies for integrating multiple knowledge systems in the context of FLR necessarily involve: (i) respectful engagement of the holders of complementary but sometimes conflicting knowledge systems; (ii) identification of legacy drivers of degradation so as to be able to mitigate threats while decolonizing current approaches to FLR that can hinder effective communication and can prevent cross-sectoral policy coordination and governance; (iii) integrating the broadly collaborative processes that often define landscape management approaches within other knowledge or management systems; and so (iv) creating processes that facilitate the opening up of Western, agency-driven models of governance to allow more collaborative and community-based approaches for real engagement (Berkes, 2007).

Traditional knowledge and its relationship to Western knowledge

Long before the introduction of 'scientific' forest management (in Europe) in the early nineteenth century and its subsequent global expansion for timber resource management (and more generally to strengthen colonial government or state control over land resources), local, often indigenous communities throughout the world managed forested landscapes relying on complex, place-based and multi-generational knowledge systems with the goal of sustaining resident communities and future generations (Berkes, 1999; Mann, 2002; Stewart, 2002; Parrotta and Trosper, 2012). The knowledge, innovations and practices of these communities evolved through detailed scientific observations and cause–effect management experiences integrated across generations. On the scale of centuries, these community-based, intergenerational stewardship systems sought to create abundance in resources that were deemed critical to community survival and livelihoods, and that were resilient and adaptive to changing environmental, economic, political and social conditions. These include, for example, indigenous cultural fire regimes (Jackson and Moore, 1998; Huffman, 2013), agroforestry practices (Walker *et al.*, 1995; McNeely and Schroth, 2006), traditional forms of watershed management (Wiersum, 1997; Mueller-Dombois, 2007) and sacred areas (Dudley *et al.*, 2009; Ormsby and Bhagwat, 2010). In altering the composition, structure, function and dynamics of forests at landscape scales (Farina, 2000; Stewart, 2002), these traditional systems purposefully altered ecological processes, resulting in enhanced abundance of desired, higher-valued species as well as the loss or reduction of some species and the extent of different ecosystem types.

The diverse knowledge systems that are embedded in the cultural traditions of regional, indigenous or local communities are referred to using a number of different terms, including traditional knowledge, local knowledge, traditional ecological knowledge, indigenous knowledge and indigenous science. In this chapter we will use a more general term, 'Traditional knowledge', defined by the Convention on Biological Diversity as 'the knowledge, innovations and practices of indigenous and local communities around the world, developed from experience gained over the centuries and adapted to the local culture and environment'. We consider Traditional knowledge in relation to Western knowledge (also called modern science, Western science or international science), that is, knowledge typically generated in universities, research institutions and private firms following paradigms and methods associated with the 'scientific method' consolidated in post-Renaissance Europe on the basis of wider and more ancient roots, usually transmitted through scientific journals, scholarly books and now internet/web-based platforms, with its central tenets being observer independence, replicable findings, systematic scepticism, and transparent research methodologies with standard units and categories (Díaz *et al.*, 2015).

Typically, Traditional knowledge is transmitted through active mentorship of younger generations, and codified orally to facilitate transmission of knowledge and associated cultural practices between generations. This oral transmission of knowledge in the case of indigenous communities takes diverse forms, including creation accounts, place names linked to socio-ecologically grounded stories, traditional law teachings associated with subsistence and ceremonial activities, all of which would be transferred through direct training of youth by elders. Such knowledge systems are strengthened by and embodied in local languages, cultural values, beliefs, rituals, stewardship practices, community laws and governance systems. The resulting observational, place-based scientific knowledge underlies a diverse array of natural resource management practices that sustain these communities' food and water security, health, cultural traditions and livelihoods (Altieri, 2002; Parrotta *et al.*, 2015; Berkes, 2017).

There are important challenges associated with, and requirements for, integration of Traditional and Western knowledge systems. Western knowledge systems often provide a more mechanistic and biophysically technical form of knowledge, whereas Traditional knowledge systems are more inclusive of holistic and metaphysical aspects. A challenge can arise when knowledge systems are pitted against each other as what is 'the best science' versus being considered complementary to enriching the collective way of knowing the landscape, or reflecting different value systems (Nadasdy, 1999; Sterling *et al.*, 2017).

While both Traditional knowledge and Western knowledge systems rely on science to understand and manage resources, they differ in philosophies and in environmental and stewardship approaches (Cajete, 2000), as summarized in Table 3.1. These important differences start with the underlying culturally determined values, norms and beliefs that influence people's perceptions and the concepts shaping their worldviews and shape their relationships to the land (Díaz *et al.*, 2015; Sterling *et al.*, 2017). Embodied in many indigenous teachings are lessons that guide humans on how to be able to 'live with their relations', which include geological (soils/landscape), biological (fungi, plants and animals) and eco-hydrological elements or processes (i.e. fire, weather, flooding, earthquakes, etc.), inclusive of all the biophysical and spiritual features contained within a landscape of interest (Lewis and Sheppard, 2005). Because Traditional knowledge systems are often place-based, they display a high degree of variation across different landscapes and forest ecosystems (Turner *et al.*, 2003). And so, the content of such knowledge fundamentally reflects integration of locally observed and codified information on a place's ecological condition and history, as well as the social, economic and cultural characteristics of the communities that have accumulated this knowledge (Berkes *et al.*, 2000; Parrotta and Trosper, 2012). These systems and associated social institutions are important components of the social capital of traditional societies, with important implications for FLR, including adaptation to environmental change (Berkes *et al.*, 2000; Galloway-McLean, 2009; Parrotta and Agnoletti, 2012; Sterling *et al.*, 2017).

Despite their diversity across regions and cultures, Traditional knowledge systems – particularly those embedded in indigenous communities – tend to share a number of common features that distinguish them from

Table 3.1 Traditional and Western knowledge approaches pertaining to forest landscape restoration

Knowledge application	Element of comparison	Traditional knowledge	Western knowledge
Philosophy	Human and resources environmental hierarchy	Kincentric: Humans and nature viewed as an extended family with shared ancestry and origins, often honoured through ritual, and a central focus of management attention.	Anthropocentric: Humans considered above all resources, where living and non-living are largely viewed through a utilitarian lens, that is, their capacity to provide ecosystem services.
Philosophy	Knowledge transfer: Teaching and gathering information	Transmitted across generations as oral traditions and cultural practices on landscape for specific habitats and resources.	Transmitted primarily through print and electronic media, in multidisciplinary academic and professional settings.
Philosophy	Data and knowledge acquisition	Information used for cultural purposes to support community wellbeing and security, use value-based condition assessments.	Information used for natural capital management, resources studied/analysed, data catalogued, fact based to assess landscape and resource condition.
Environment	Environmental phenomena	Biophysical explanations are complemented by metaphysical explanations for observed phenomena, including extreme events.	Biophysical explanations of environmental phenomena. Natural range of variation concept applied as explanation for extreme or unusual events.
Environment	Resolution or accuracy of data	Quantitative and qualitative: Assessments determined by past and current cultural use to predict future conditions or amounts. Patterns are anticipated based on complex phenological indicators and past experiences. Local focus.	Quantitative and qualitative: Assessments by field data, modelling, analysis and trends predicted. Patterns are anticipated based on complex data collecting methods and incorporation into simulation models. Local to global focus.

Category	Attribute	Renewal	Recovery
Environment	Use of reference ecosystems to guide FLR restoration and management actions	Renewal: Human services to ecosystems. Past social-ecological systems, where appropriate, guide current and future actions. Fewer technological fixes as solutions.	Recovery: Ecosystem services to humans. Past biophysical conditions or novel ecosystems. Reliance on technological solutions and regulatory compliance to control environment and influence landscape processes.
Stewardship	Values or obligation to environment	Socio-cultural and spiritual responsibility to landscape. Management is self-regulated via cultural norms. Negative impacts to valued resources are avoided, adhering to beliefs and tenets.	Social and economic responsibility to private and public stakeholders. Negative impacts to valued resources are spatially and temporally mitigable through existing policies, governance institutions and management practices.
Stewardship	Environmental management	Ethically based: Holistic stewardship a spiritual obligation and sacred responsibility for resource use. Humans necessary part of landscape.	Socio-economically based: Driven by policies, markets, laws and/or regulations. Sustainability based on thresholds of resource use. Humans not necessarily an integral part of landscape.
Stewardship	Adaptive learning/ principles	Learned from ancestral teachings (passed on knowledge), observation and direct experiences with landscapes, habitats and resources. Corrective actions to human uses to benefit resources.	Field observations, scientific literature inform conceptual theories and principles taught and applied to management of habitats and other biophysical resources. Corrective actions on resource use to benefit humans.

Western knowledge systems and associated natural resource management practices (Berkes *et al.*, 2000; Cajete, 2000; Trosper *et al.*, 2012; Donatuto *et al.*, 2014). These include the high value placed on:

1   Sustainability: retaining or enhancing the ecological, social, economic, cultural and spiritual values of the land;
2   Relationships: people's connections among themselves and to their territory are not severed by the use of new knowledge, ideas or techniques;
3   Identity: communities seek to maintain their distinct cultural identities;
4   Reciprocity: people maintain their system of benefit sharing among members of the community;
5   Limitations placed on market involvement: while people may engage in market exchange with the flow of goods and services from the land, the fundamental productivity of the system itself is not viewed as a resource to be exchanged.

These values help to explain how, in the absence of external and internal pressures that result in erosion or destruction of traditional cultural and spiritual values and governance institutions, or loss of connection to their lands, Traditional knowledge and practices have survived, evolved and sustained local and indigenous communities over generations through changing environmental and socio-political conditions (Lewis and Sheppard, 2005).

As illustrated in Table 3.2 (adapted from Berkes *et al.*, 2000), traditional knowledge systems encompass a broad array of land management practices as well as underlying social mechanisms that facilitate the development, sharing and intergenerational transmission of knowledge, the functioning of local institutions affecting land management, and mechanisms for maintaining and reinforcing shared cultural values.

These systems can include many, if not most, of the necessary elements for success in forest landscape restoration: management practices for maintaining or enhancing biodiversity and provision of a range of ecosystem services; the necessary social institutions required for developing shared visions of FLR aims/objectives; strategies for reconciling the needs of diverse stakeholders; and approaches for adaptive management and sharing of knowledge as well as the risks and benefits of FLR implementation (Brown, 2005; O'Connor *et al.*, 2005; Boedhihartono and Sayer, 2012; Sterling *et al.*, 2017).

Integration across knowledge systems

In the same way that there is increasing recognition of the value of considering both social systems and ecological systems together, in what are termed 'social-ecological systems' (Chapter 5, this volume), there is also an opportunity to integrate Traditional knowledge and Western knowledge in

Table 3.2 Social-ecological practices and mechanisms in Traditional knowledge and practices.

Management practices based on ecological knowledge	Social mechanisms behind management practices
Practices found both in conventional resource management and in some local and traditional societies: • Monitoring resource abundance and change in ecosystems; • Total protection of certain species; • Protection of vulnerable life history stages	*Generation, accumulation and transmission of local ecological knowledge:* • Reinterpreting signals for learning; • Revival of local knowledge; • Folklore and knowledge carriers; • Integration of knowledge; • Intergenerational transmission of knowledge; • Geographical diffusion of knowledge
Practices largely abandoned by conventional resource management but still found in some local and traditional societies: • Multiple species management; • Maintaining ecosystem structure and function; • Resource rotation; • Succession management	*Structure and dynamics of institutions:* • Roles of stewards/wise people; • Cross-scale institutions; • Community assessments; • Taboos and regulations; • Social and religious sanctions
Practices related to the dynamics of complex systems, seldom found in conventional resource management but found in traditional societies: • Management of landscape patchiness; • Watershed-based management; • Managing ecological processes at multiple scales; • Responding to and managing pulses and surprises; • Nurturing sources of ecosystem renewal	*Mechanisms for cultural internalization:* • Rituals, ceremonies, and other traditions; • Cultural frameworks for resource management *Worldview and cultural values:* • A worldview that provides appropriate environmental ethics/ tenets; • Cultural values of respect, sharing reciprocity, humility and other

Source: adapted from Berkes *et al.* (2000); used with permission.

environmental conservation and restoration. In practice, this means understanding the former legacy of or current desires for indigenous landscape forest management, while respectfully challenging current scientific methods, approaches and beliefs about socio-economic institutions of forest-dependent cultures. It requires adopting new understandings of the landscape in question and its stakeholders, and adopting or designing new tools that can align application for both sets of knowledge systems.

In this book, our intention is to challenge the current 'uni-dimensional' ways of approaching FLR. We recognize the shortcomings of narrow disciplinary and sectoral approaches, and the need for a broader perspective on FLR that considers integration across scales, across disciplines and across knowledge systems. Chapter 12 later in this volume focuses on the value of integrating knowledge systems for effective FLR implementation.

References

Altieri, M.A. (2002) 'Agroecology: The science of natural resource management for poor farmers in marginal environments', *Agriculture, Ecosystems and Environment*, vol 93, pp. 1–24.

Berkes, F. (1999) *Sacred ecology: Traditional ecological knowledge and management systems*. Taylor & Francis, Philadelphia and London.

Berkes, F. (2007) 'Community-based conservation in a globalized world', *Proceedings of the National Academy of Sciences*, vol 104, pp. 15188–15193.

Berkes, F. (2017) *Sacred ecology*, 4th ed. Routledge, New York.

Berkes, F., Colding, J. and Folke, C. (2000) 'Rediscovery of traditional ecological knowledge as adaptive management', *Ecological Applications*, vol 10, pp. 1251–1262.

Boedhihartono, A.K. and Sayer, J. (2012) Forest landscape restoration: Restoring what and for whom? In Stanturf, J., Lamb, D. and Madsen, P., eds. *Forest landscape restoration* (pp. 309–323). Springer, Dordrecht.

Brown, K. (2005) Addressing trade-offs in forest landscape restoration. In Mansourian, S., Dudley, N. and Vallauri, D., eds. *Forest restoration in landscapes* (pp. 59–64). Springer, New York.

Cajete, G. (2000) *Native science: Natural laws of interdependence*. Clear Light Press, Santa Fe, NM.

Díaz, S., Demissew, S., Carabias, J., Joly, C., Lonsdale, M., Ash, N., Larigauderie, A., Adhikari, J.R., Arico, S., Báldi, A. and Bartuska, A. (2015) 'The IPBES Conceptual Framework – connecting nature and people', *Current Opinion in Environmental Sustainability*, vol 14, pp. 1–16.

Donatuto, J., Grossman, E.E., Konovsky, J., Grossman, S. and Campbell, L.W. (2014) 'Indigenous community health and climate change: Integrating biophysical and social science indicators', *Coastal Management*, vol 42, no 4, pp. 355–373.

Dudley, N., Higgins-Zogib, L. and Mansourian, S. (2009) 'The links between protected areas, faiths, and sacred natural sites', *Conservation Biology*, vol 23, no 3, pp. 568–577.

Farina, A. (2000) 'The cultural landscape as a model for the integration of ecology and economics', *BioScience*, vol 50, no 4, pp. 313–320.

Galloway-McLean, K. (2009) *Advance guard: Climate change impacts, adaptation, mitigation and indigenous peoples – a compendium of case studies*. United Nations University-Traditional Knowledge Initiative, Darwin, Australia.

Huffman, M. (2013) 'The many elements of traditional fire knowledge: Synthesis, classification, and aids to cross-cultural problem solving in fire-dependent systems around the world', *Ecology and Society*, vol 18, no 4, article 3. Available at: http://dx.doi.org/10.5751/ES-05843-180403 (accessed 19 June 2018).

Jackson, W.J. and Moore, P.F. (1998) The role of indigenous use of fire in forest management and conservation. In: *International seminar on cultivating forests:*

Alternative forest management practices and techniques for community forestry. Regional Community Forestry Training Center, Bangkok, Thailand.

Lewis, J.L. and Sheppard, S.R. (2005) 'Ancient values, new challenges: Indigenous spiritual perceptions of landscapes and forest management', *Society and Natural Resources*, vol 18, no 10, pp. 907–920.

Mann, C.C. (2002) '1491', *The Atlantic Monthly*, March, 2002 (13 p.).

McNeely, J.A. and Schroth, G. (2006) 'Agroforestry and biodiversity conservation – traditional practices, present dynamics, and lessons for the future', *Biodiversity & Conservation*, vol 15, no 2, pp. 549–554.

Mueller-Dombois, D. (2007) 'The Hawaiian ahupua'a land use system: Its biological resource zones and the challenge for silvicultural restoration', *Biology of Hawaiian Streams and Estuaries. Bishop Museum Bulletin in Cultural and Environmental Studies*, vol 3, pp. 23–33.

Nadasdy, P. (1999) 'The politics of TEK: Power and the "integration" of knowledge', *Arctic Anthropology*, pp. 1–18.

O'Connor, S., Salafsky, N. and Salzer, D. (2005) Monitoring forest restoration projects in the context of an adaptive management cycle. In Mansourian, S., Dudley, N. and Vallauri, D., eds. *Forest restoration in landscapes* (pp. 145–149). Springer, New York.

Ormsby, A.A. and Bhagwat, S.A. (2010) 'Sacred forests of India: A strong tradition of community-based natural resource management', *Environmental Conservation*, vol 37, no 3, pp. 320–326.

Parrotta, J.A. and Agnoletti, M. (2012) Traditional forest-related knowledge and climate change. In Parrotta, J.A. and Trosper, R.L., eds. *Traditional forest-related knowledge: Sustaining communities, ecosystems and biocultural diversity* (Chapter 13, pp. 491–533). Springer, Dordrecht.

Parrotta, J.A. and Trosper, R.L. (eds) (2012) *Traditional forest-related knowledge: Sustaining communities, ecosystems and biocultural diversity*. World Forest Series vol. 12. Springer, Dordrecht, the Netherlands.

Parrotta, J.A., Dey de Pryck, J., Darko Obiri, B., Padoch, C., Powell, B. and Sandbrook, C. (2015) The historical, environmental and socio-economic context of forests and tree-based systems for food security and nutrition. In Vira, B., Mansourian, S. and Wildburger, C., eds. *Forests and food: Addressing hunger and nutrition across sustainable landscapes* (Chapter 3, pp. 71–134). Open Book Publishers, Cambridge, UK. http://dx.doi.org/10.11647/OBP.0085.

Ruiz-Mallén, I. and Corbera, E. (2013) 'Community-based conservation and traditional ecological knowledge: Implications for social-ecological resilience', *Ecology and Society*, vol 18, no 4, p. 12.

Sterling, E.J., Filardi, C., Toomey, A., Sigouin, A., Betley, E., Gazit, N., Newell, J., Albert, S., Alvira, D., Bergamini, N. and Blair, M. (2017) 'Biocultural approaches to well-being and sustainability indicators across scales', *Nature Ecology & Evolution*, vol 1, no 12, p. 1798.

Stewart, O. (2002) *Forgotten fires: Native Americans and the transient wilderness*. University of Oklahoma Press, Oklahoma.

Trosper, R.L., Parrotta, J.A., Agnoletti, M., Bocharnikov, V., Feary, S.A., Gabay, M., Gamborg, C., Latorre, J.G., Johann, E., Laletin, A., Fui, L.H., Oteng-Yeboah, A., Pinedo-Vasquez, M., Ramakrishnan, P.S. and Youn, Y.C. (2012) The unique character of traditional forest-related knowledge: Threats and challenges ahead. In Parrotta, J.A. and Trosper, R.L., eds. *Traditional forest-related knowledge: Sustaining communities, ecosystems and biocultural diversity* (pp. 563–588). Springer, Dordrecht.

Turner, N.J., Davidson-Hunt, I.J. and O'Flaherty, M. (2003) 'Living on the edge: Ecological and cultural edges as sources of diversity for social-ecological resilience', *Human Ecology*, vol 31, no 3, pp. 439–463.

Walker, D.H., Sinclair, F.L. and Thapa, B. (1995) Incorporation of indigenous knowledge and perspectives in agroforestry development. In Sinclair, F.L., ed., *Agroforestry: Science, policy and practice*, vol. 47 (pp. 235–248). Springer, Dordrecht.

Wiersum, K.F. (1997) 'Indigenous exploitation and management of tropical forest resources: An evolutionary continuum in forest-people interactions', *Agriculture, Ecosystems & Environment*, vol 63, no 1, pp. 1–16.

4 Power, inequality and rights

A political ecology of forest restoration

Nitin D. Rai, Suhas Bhasme and Poorna Balaji

Introduction

Forest management has a long and contested history. Global efforts to address environmental issues have often identified the conservation of forests as a major solution to biodiversity loss and environmental decline. The interest in the conservation of forests in many tropical regions follows from a historical legacy of European colonial regimes using and managing forests as engines of economic production. Such production systems required wresting control over forests from local people and institutions. As forests were central to the state's economic production, a vast forest administration was established for the management and conservation of forest lands. Local people whose use of forest conflicted with state objectives were blamed for the decrease in forest cover (see Fairhead and Leach, 1996 for West Africa; Williams, 2006 for India; and Scales, 2011 for Madagascar). The narrative that these regions were once forested and subsequently degraded due to local use was used to restrict local use and centralize control. In some places, timber extraction was eventually followed by large-scale reforestation, while in others, exploitation was followed by conversion to other land uses. In countries such as India, country-wide targets were set for the area of land that should be maintained under forest through protection or restoration. In many instances, the social outcomes of forest extraction and subsequent reforestation efforts have been adverse. In recent times, the realization that forests could play a role in climate mitigation has created a different commodity – carbon – that could be traded to mitigate fossil fuel emissions. In this way, forests continue to serve as a global commodity even as they play a major role in ensuring the livelihoods of local people.

In recent times, the influence of global narratives about the importance of forests as carbon sinks and refuges for biodiversity has heightened the desire by states to meet national and/or internationally determined targets. A global movement, forest landscape restoration (FLR), has now been initiated for the restoration of degraded lands. As an approach, FLR is seeing widespread dissemination, notably by The Global Partnership on Forest

Landscape Restoration (GPFLR), a consortium of 25 government and non-governmental organizations (NGOs), which has so far implemented projects in about 17 countries across different cultures, geographies and ecologies, and the Bonn Challenge, which has obtained commitments towards FLR from 45 countries to date (see: www.bonnchallenge.org). The GPFLR estimates that more than 2 billion hectares of the world's deforested and degraded landscapes have potential for restoration, a process that they claim would help reduce poverty, improve food security, mitigate climate change, conserve biodiversity, improve soil and water protection, and increase the world's forest area from 31% to 47% (Laestadius *et al.*, 2011). The United Nations Food and Agriculture Organization (FAO) launched a Forest Landscape Restoration Mechanism to help countries to meet the Bonn Challenge to restore 150 million hectares of degraded and deforested lands by 2020 (Laestadius *et al.*, 2011). These are major interventions that are bound to have big outcomes for landscapes and people. The extent and scale at which reforestation efforts are now being implemented, therefore, requires us to pay particular attention to the inequalities and conflicts that are produced by large-scale reforestation and afforestation programmes.

The concept of FLR was initially developed as an alternative to more traditional approaches, which have often been site-based, focusing on one or a few forest products, relying heavily on tree planting of a limited number of species, and failing to address the root causes of forest loss and degradation (Newton and Tejedor, 2011). FLR attempts to fill these gaps by 'regaining ecological integrity and enhancing human well-being in deforested and degraded landscapes' (Dudley *et al.*, 2005). Much of the scholarship thus far perceives FLR as providing a collaborative platform for governments, NGOs and other private sector organizations to develop a range of solutions to tackle a multifaceted problem such as forest degradation (Bekele-Tesemma and Ababa, 2002). Unlike traditional strategies of agroforestry and forest restoration focusing on a particular section of agricultural land or forest area, FLR targets the landscape using techniques and methods to improve the condition of these degraded lands.

As the approach is targeted at the landscape level, there is a risk that it could create differences between stakeholders, and affect landscape access and use by certain sections of the population, thereby creating or exacerbating inequalities. These inequalities may occur within the landscape, or between the landscape and other scales (national, local, international). They may occur across groups (caste, gender, ethnicity) within the landscape, or between groups at different scales (e.g. private multinational companies vs. local community), mediated by power dynamics.

The idea of restoration itself might be contested by different stakeholders, including national and international actors involved in the implementation of the programme. The notion of forest landscape degradation and then of restoration is a matter of perception, depending on one's social and economic location (Chazdon *et al.*, 2016). There is no precise definition of what constitutes forest and forest degradation across the globe at either national or international level. The Food and Agriculture Organization (FAO) defines forest as

> land spanning more than 0.5 hectares with trees higher than 5 meters and a canopy cover of more than 10 percent, or trees able to reach these thresholds in situ. It does not include land that is predominantly under agricultural or urban land use.
>
> (FAO, 2015)

Implementing agencies (mostly national and international), who might use the standard definition adopted by the UN, about what constitutes a forest may find that this definition conflicts with local people's idea of what a forest is (Chazdon *et al.*, 2016).

In light of studies that discuss the impact of different perceptions of forest condition, we believe that the idea of restoration is contestable among the local, regional, national and international actors working on the FLR agenda. Local people might perceive the area and condition of forest to have improved, while external actors – such as national and international agencies prescribing to a different definition – might see forests as degrading, based on narratives of regional forest loss used to justify wresting control of forests from local people (Fairhead and Leach, 1996). These perceptual differences in forest extent and condition lead to confusion and often conflict between local people and the programme-implementing agency. Forests are in continuous transition and influenced by factors such as agricultural development, urbanization, industrialization and migration. In this chapter, we address the conflicts and inequalities that arise from the efforts to reforest landscapes and highlight ways in which restoration projects might address them.

The political ecology of forest restoration

We use political ecology to explore the linkages and outcomes of larger global environmental policy on local social and political dynamics. It enables us to highlight the broader social and political inequalities existing among the stakeholders involved in such conservation programmes implemented in developing countries. Political ecology also allows us to address the inequalities that exist at several levels: at the national and regional levels, as well as within local groups.

To understand the opportunities and constraints of FLR, it is necessary to identify the contexts in which inequalities exist and the relationships of power among stakeholders. A political ecology approach can also help to understand the larger political economy of environmental conservation within which FLR takes place, along with the challenges involved in the process of implementation at the local level. What is the institutional setting? What are the different stages of implementation? Who are the actors in each level of implementation governance? How do they interact with one another? How does it account for local politics? Political ecology provides the space to engage with these questions, pay attention to multiple scales, and tease out both social and ecological inequalities. Mainly, it allows us to question the basic rationale driving FLR, and to understand how it has framed local actors within its ambit, as they are the ones who are deeply implicated on the ground – as either winners or losers. By identifying such winners and losers, it offers the possibility for the implementers of FLR to offer corrective actions to adverse outcomes as they occur. Political ecology could be the 'hatchet' as well as the 'seed' in analysing environmental projects that have possible implications for people (Robbins, 2011).

Political ecology is about not only politicizing the environmental issues but also 'ecologizing' the political process (Bryant and Bailey, 1997). The approach is explicitly about understanding ecological processes in their social and political contexts, involving issues of power and hierarchies. Scholars have used a political ecology approach to challenge dominant Malthusian narratives of the relationship between population growth and degradation of the environment. They have offered a more nuanced explanation for some of the grave environmental problems of the times, including the famine in the Sahel and growing food crises in Sub-Saharan Africa (Franke and Chasin, 1980), the degradation of Himalayan forests (Forsyth, 1996) and land degradation in East Africa (Tiffen et al., 1994). These explanations of environmental problems explicitly highlighted the linkages between the international economy and local political and social processes in the affected regions. In doing so, they reveal the ways in which social categories such as caste, class, gender and ethnicity affect, and are affected by, environmental issues.

There is a growing motivation for governments to counter the deleterious consequences of dominant economic production models by adopting international conservation discourses within national conservation policies. Notably, the growing influence of REDD+ (reducing emissions from deforestation and forest degradation, and the role of conservation, sustainable management of forests, and enhancement of forest carbon stocks in developing countries) and reforestation initiatives under the United Nations Framework Convention on Climate Change's Nationally-Determined Contributions (NDCs) have led to the accelerated implementation of forest restoration initiatives. The following sections discuss the inequality that exists among stakeholders at several levels – at the national and regional levels, as well as within local groups – and their implications for FLR initiatives. While we consider the literature and experiences from throughout the world, our particular focus is on India, where a number of lessons have been learned that can inform FLR initiatives worldwide.

The influence of national policies on local actors

In the last two decades, there has been a massive socio-economic and political transformation in the Global South. In terms of class domination, these changes, in the case of India, for example, can be summed up as a transformation from the earlier three dominant proprietary classes – industrial capitalists, rich farmers and the salariat (Bardhan, 1984) – to the current domination of industrial capitalists within the state apparatus (Chatterjee, 2008). Simultaneously, however, the uprising of socially marginalized groups, particularly due to identity politics of lower caste groups and large environment movements, has led to the deepening of Indian democracy (Baviskar, 1999; Jaffrelot, 2003; Gupta and Sivaramakrishnan, 2016). This transformation has changed the relationship between the state and society via decentralization, building up local institutions, and transferring authority to local people (Baviskar and Matthew, 2009).

Understanding the role of the state and its policies is important for evaluating the likely or actual on-the-ground consequences of an intervention such as FLR that is conceived as a collaborative effort that includes governments, NGOs and other private organizations (Bekele-Tesemma and Ababa, 2002). National policies determine several aspects of forest management, including definitions, nature of tenure, and territorialization of forests required to meet conservation goals (Reyes-García *et al.*, 2014; Chazdon *et al.*, 2016). Definitions of forests as developed by the state provide the conceptual, institutional, legal and operational basis for the policies and monitoring systems that drive or enable deforestation, forest degradation, reforestation, and forest restoration (van Noordwijk and Minang, 2009; Chazdon *et al.*, 2016). State interventions are influenced by international discourses on conservation but rarely by discourses from the ground. Local actors rarely play a role in shaping policies, even though they may be directly implicated. Even initiatives such as Joint Forest Management in India were initiated and controlled by state and international players and not by grassroots movements that were campaigning for local rights to forests. Creation of national policies has always been skewed towards state actors.

The effect of reforestation on local rights and tenure

FLR's landscape approach considers forest patches within the larger spatial matrix of non-forest land uses and also incorporates the multiple functions of forests in order to address livelihood objectives. This means that the influence of national policies in implementing FLR activities will go beyond those of forest management but must take into account all the other non-forest activities in the area. Existing state policies may play a pivotal role in the design of FLR activities, and there may be a danger of these activities exacerbating inequalities already in place due to state interventions, particularly where national policies are formulated in a centralized manner with no representation from the communities that are affected by these policies.

The case of the Compensatory Afforestation Fund Management and Planning Authority Act (CAMPA) in India is a good example of an afforestation programme whose implementation has affected not just the rights of local people but also tenurial arrangements due to the appropriation of common lands for plantation. The Compensatory Afforestation Fund Act was enacted in 2016 as a broad guideline to utilize the funds collected from the user agencies, such as mining companies, to offset forest conversion. The revenues from these offsets, which were collected from industries that deforested an area of land, have been used to plant trees in forest and non-forest areas. These areas, which the state identifies as 'degraded' forest, are often used by local people for a range of activities including cultivation, grazing, and collection of plants and other products. Some authors have even called such an appropriation of areas for plantation a 'land grab' (Karthik and Kodiveri, 2018). Studies on CAMPA across the country have shown that the process is centralized and top-down, with little space for negotiations with local communities (Kohli and Menon, 2011). From identification of land for compensatory afforestation (CA) to selection of species to be planted, decisions are made by the Forest Department. In most CA sites, the Forest Department plants teak, which is classified by the department as an economic species, much against the wishes of local communities, who would have preferred fruit trees. Activities carried out under CA affect tenure security, as the state does not recognize de facto rights, and many of the plantations are undertaken on shifting cultivation land, thereby affecting the livelihoods of these farmers (Temper and Martinez-Alier, 2013; Sahu *et al.*, 2017).

Villages are often comprised of heterogeneous communities with both landless and landed families, and past experiences of implementing integrated conservation and development programmes have produced complex outcomes, which have often benefitted landholders and left landless people, including women, particularly vulnerable (Vira *et al.*, 2012). As this example illustrates, when FLR initiatives are undertaken in different countries, programme implementers may need to consider inequalities perpetrated by existing national policies.

The impact of international policies: lessons from REDD+

Forest restoration is a multi-level process that involves international funding agencies. In the recent past, with the advent of valuation of ecosystem services and the introduction of Payment for Ecosystem Services (PES) schemes, there has been renewed interest by multinational enterprises in restoration as a means to maintain their industrial production by compensating communities or states in distant regions to protect and regenerate forests (Büscher, 2012). Under such schemes, companies situated in developed countries may compensate for their carbon-intensive activities through interventions in the form of forest restoration in developing economies. International companies are helped by state and international agencies, which set goals and policies for forest restoration. By being singularly focused on carbon sequestration, reforestation programmes tend to exclude local aims and perceptions. In addition to the marginalization of local people, even options that may be more environmentally and socially beneficial, such as natural regeneration and the planting of indigenous species, are frequently ignored in forest restoration projects (McElwee, 2009).

One prominent and emerging initiative that has much international backing but has the potential for local conflict is REDD+, an international climate change mitigation approach that provides financial benefits to countries for reducing greenhouse gas emissions by avoiding degradation and loss of forest (Agrawal *et al.*, 2011). REDD+ involves a range of stakeholders, from local forest dwellers to regional, national and international actors. At present, countries engaging in REDD+ are facing stiff challenges in implementing the programme and delivering its benefits to local communities, often with adverse outcomes (Norman and Nakhooda, 2015; Svarstad and Benjaminsen, 2017). In the following sections, we discuss some of these challenges affecting the implementation of the programme, specifically those related to land tenure, reconciliation of development versus environment priorities, community participation, and increased power of state actors. These experiences from REDD+ implementation provide relevant lessons for developing a more inclusive approach to FLR.

Land tenure

In many parts of the forested tropics, landless people often practise traditional agriculture in the form of shifting cultivation, enjoying de facto rights over forest land. The lack of secure land tenure of these forest-dependent people often increases the risk of their displacement by the state under the agenda of development or conservation of forests and wildlife. A comparative study of the implementation of REDD+ points out the complications raised by the lack of secure land rights of forest-dependent people (Larson *et al.*, 2013). Another study of 19 REDD+ project sites located in Brazil, Cameroon, Tanzania, Indonesia and Vietnam points out that without formal recognition of land rights, communities lack support and authority to prevent incursion of 'outsiders' into forests (Sunderlin *et al.*, 2014). These outsiders, mainly loggers and private timber industry, tend to easily override the customary rights of forest-dependent people lacking legal support from the state. In the absence of secure land tenure in the case of countries like Peru, Indonesia and Cameroon, there is continuing conflict over land rights between the state and the community (Bayrak and Marafa, 2016).

Community participation

Community participation is another aspect that needs consideration for effective implementation of REDD+ (Agrawal and Angelsen, 2009). A policy note from the World Bank on REDD+ emphasizes the importance of free, prior and informed consent of the participants, monitoring and accounting of carbon stocks, local participation in reforestation, and other activities for effective implementation (World Bank, 2011). However, studies on the implementation of REDD+ show that safeguards surrounding free, prior and informed consent, which is part of the UN-REDD policy framework (Peskett and Todd, 2012), are violated or given only symbolic importance by implementing authorities (Bayrak and Marafa, 2016; Svarstad and Benjaminsen, 2017). In the case of Vietnam, one of the leading countries implementing a REDD+-readiness plan, it is observed that local people were ignorant about the programme and its potential benefits, and that the 'community consent' was obtained by the state officials without any serious discussion with local people (Tan et al., 2010). In other cases, such as in Nepal, communities report the loss of autonomy following the implementation of project guidelines (Poudel et al., 2014). Lund et al. (2017) examined Tanzanian REDD+ policy and practice and suggested that even as REDD+ offers a 'discursive change' in forest governance, it 'delivers continuity' in state forest control regimes. Highlighting the increasing marginality of local voices in carbon forestry projects, Beymer-Farris and Bassett (2012) warn that local people might show increasing resentment and resistance to REDD+ due to fears that they might be made vulnerable and even displaced. Proponents of REDD+ and other carbon forestry projects should, therefore, actively engage with environmental justice concepts to ensure more equitable and participatory outcomes of these climate mitigation initiatives. The question has been raised whether REDD+ and similar reforestation efforts might result in the recentralization of forest governance (Phelps et al., 2010).

Increasing power of the state

The challenge of empowering local people *vis à vis* the state should be seen as a recurring issue, particularly in the Global South, and one not restricted to REDD+. Most forest programmes implemented in these regions, such as social forestry and Joint Forest Management (JFM), that advocate the empowerment of local communities face stiff challenges and resistance from the state and local elites (Sundar et al., 2001; Lele and Menon, 2014). State Forest Department actions in the case of India are guided by 'self-regarding behaviour' and 'professional logic of appropriateness' that negates the voice of forest-dependent people (Fleischman, 2014). For example, in 1952 the Indian government set a goal of 33% of India's land area to be under forest cover (Joshi *et al.*, 2011). This goal has now been reinforced under a major national effort called the Green India Mission, which hopes to achieve the 33% targets (Robbins and Tripuraneni, 2017). Such state efforts to achieve a country-wide increase in forest cover are sure to have impacts on people who are dependent on existing forest areas for their livelihoods (Sarin, 2005) as well as on the composition of the forest (Martin, 2003).

FLR programmes need to acknowledge the challenges faced by REDD+ and attempt to move beyond the narrow rationale of reforestation of a landscape as a panacea for addressing climate change and global goals. A democratic and equitable FLR approach should address the existing power relations among the state, civil society groups, national and international agencies, business enterprises, and local communities, as well as within communities.

Gender and inequality

Marginalized people the world over, and particularly women in the Global South, face major livelihood threats due to climate change and forest degradation (Robertson and Lawes, 2005; Skinner, 2011). Due to their lack of secure land rights, women are often more dependent on forests for their subsistence in the form of collection of non-timber forest products such as fruits and medicinal plants, fuelwood and food (Agarwal, 1997). For the last three decades, efforts have been made by the state to share its responsibility to manage forest with communities and include women in the governance of forest, particularly in the Global South. The issue of 'gender matters' raised by Third World scholars such as Shiva (1988) and Agarwal (2001) has striven to establish a strong link between women and the environment. Agarwal (1994, 1997) points out that gender division of resource-based labour and culturally specific gender roles led to the formation of a 'special relationship' between women and the environment.

Any initiative to manage forests at the local level should include women because of their everyday dependency on forests and situated knowledge about forests. It is argued that the inclusion of women in community management of forests will enable equity (Agarwal, 1997, 2001). The gender dimension in forest management led to the formulation of the idea commonly referred to as 'Women, Environment and Development' (WED), which highlights the 'special link' between women and the environment, and emphasizes the need to include women in all activities concerning forest management. However, women do not necessarily represent a homogeneous group but, rather, one that can be further differentiated on the base of ethnicity, religion, caste and class. These social and economic distinctions lead to differences regarding opportunities, participation and experience of the public sphere among women (Fortmann and Bruce, 1991). Any participatory approach that seeks to address gender issues should consider these social and economic differences and their effects on the level of women's participation in collective action initiatives.

With this in mind, we discuss two prominent positions on women's participation in forest management. The first one argues for the inclusion of women in forest governance due to their proximity to the forest and knowledge of local conditions. An alternative position questions the essentializing role of women as caretakers of nature and enquires more about how structural limitations and changing economic scenarios influence the ability and will of women to participate in forest management. The discussion on these two positions is informed by studies that examine the ways in which women's participation is hampered by local-level organizations. Finally, we explain why a participatory approach aimed at empowering women needs to critically look at the woman–environment link in a given context, and understand women as individuals located in a larger scenario of development, who may or may not choose to participate in forest management.

Women, participation and community forest groups

Despite efforts to highlight the role of women in forest management, women have often been excluded from the organization of local groups to manage forests. In an analysis of gender participation in community forest management in South Asia, Agarwal (2001) points out that women are excluded from most state-initiated local community forest groups (CFGs). Based on a study of 87 CFGs located across India and Nepal, Agarwal (2001) shows that even though membership in the 'General Body' of the CGF is given to each household in the village, it is generally the men from these households who participate in management decisions and activities, with women forming less than 10% of the total membership of the organization. The low proportion of membership of women also affects their representation in the decision-making bodies, which remain largely male-centric, with power largely in the hands of men of dominant caste or class, marginalizing the voices of the poor and women (Chhatre, 2007; Cornwall, 2007). The exclusion at the organizational level eventually makes women passive members of the organization, dependent entirely on varied sources of information: husbands, male members of the family and limited local women's networks. This exclusion of women is not restricted only to state-initiated groups but is also observed in informal traditional groups formed by local communities in many parts of India and elsewhere in the Global South (Meinzen-Dick et al., 1997; Agrawal, 2005).

However, even in the case of groups formed by women, it has been observed that gender differentiation at a societal level continues to affect the functioning and effectiveness of these groups, such as with regard to access to knowledge about new technology and ways to acquire it. Mwangi and co-authors (Mwangi *et al.*, 2011) show that women's groups are excluded from access to technology that offsets pressure on forests and increases the income of the members through activities like bee-keeping and seedling planting. Men who have traditionally been in the business of forest products tend to have good relations with (predominantly male) forest extension officials, who provide information about the different programmes and state schemes. It is evident that in a given context, the forest extension official will either hesitate or not share information with women's groups. The other important factor that affects the access of women to technology is an economic hierarchy at the household level that gives women limited control over cash to purchase such livelihood-enhancing inputs as bee houses or seeds.

Cultural and structural issues hampering women's participation

The social perceptions and cultural roles of women can often restrict their access to knowledge and ability to procure the means to manage the forest. Merely paying lip service to enhanced participation at an organizational level without addressing the gender differentiation at the social level might not lead to desired goals of women's empowerment or forest conservation. Instead, the more substantial structural inequalities that shape as well as sustain the gender bias in forest governance need to be addressed (Leach, 2007; Resurreccion and Elmhirst, 2012). The analysis of gender issues with relation to forest management typically ignores the structural basis of this gender division and in one way or another represents 'gross essentialism' with 'patronising parental' thinking central to development discourse (Cornwall, 2007). The argument of 'women for the environment' may only reinforce existing social inequalities in the social roles of men and women, thereby exacerbating differences with no environmental benefit (Rankin, 2001). It has also been argued that as a result of women empowerment programmes, the organizational responsibility of participation may place a further burden on women, who struggle to perform their duties as mothers and wives at a household level (Datta, 1998; Cornwall and Rivas, 2015).

Situating women within a changing political economy

Similarly, the 'women–nature link' ignores the process of economic liberalization that provides rural populations, including women, with a wide range of opportunities to diversify their livelihoods in the emerging economies in Asia and Africa (Razavi, 2002; Carswell and De Neve, 2013). In light of newly available opportunities, women might withdraw from their traditional role of resource managers not merely out of economic interest but also out of the need to claim a new identity in the process of development (Resurreccion, 2006). Depending on their economic or social needs, women might position themselves differently in relation to forest and state- or community-sponsored forest management programmes.

All in all, any gender-based approach needs to be cautious while addressing gender disparity and women's participation in forest management. It is important to have historical as well as social knowledge about gender disparities, but more importantly, to understand the context in which the programme is being implemented. A 'one size fits all' approach might turn out to be more problematic than no plan at all!

Conclusion

The historical appropriation of forests by state and non-local agencies in many countries has led to loss of local rights as well as forest degradation and loss due to timber extraction and land use conversion. There has, therefore, been a perceived need for the restoration of forests that were impacted by a skewed developmental agenda. And yet, there is considerable evidence that past forest restoration programmes have had equally adverse outcomes on local people due to the non-participatory nature of the interventions and the perceptual differences between the state and local people on the meaning and role of forests (McElwee, 2012). Reforestation is not merely a local and physical intervention but is often tied to larger political economic contexts and drivers. The implementation of forest restoration requires consideration of a long chain of processes and strategies.

While forests have long been appropriated for such developmental purposes as mining, dams and infrastructure, more recently a different sort of appropriation has been recognized, which has been referred to as a 'green grab'. Such interventions include protected areas and large-scale tree plantation projects, ostensibly to enhance conservation. Fairhead *et al.* (2012) define green grab as 'the appropriation of land and resources for environmental ends'. This is being actively pursued in the most recent phase of commodification of nature through valuation of the ecosystem services provided by forests. Such an approach emphasizes one particular relationship with nature while ignoring the cultural and social relationships between people and forests. The case of CAMPA in India, and REDD+ in a number of other countries, shows that the marginalization of local people by more powerful actors is due to differing interests and motivations. Forest restoration has the potential to exacerbate existing inequalities unless these power dynamics are acknowledged and dealt with appropriately. FLR projects need to consider these inequalities among the different stakeholders for a more inclusive implementation, however difficult this might be. A political ecology approach helps implementers take note of the challenges and plan strategies that recognize these challenges.

References

Agarwal, B. (1994) *A field of one's own: Gender and land rights in South Asia* (Vol. 58). Cambridge University Press, Cambridge.

Agarwal, B. (1997) ' "Bargaining" and gender relations: Within and beyond the household', *Feminist Economics*, vol 3, no 1, pp. 1–51.

Agarwal, B. (2001) 'Participatory exclusions, community forestry, and gender: An analysis for South Asia and a conceptual framework', *World Development*, vol 29, no 10, pp. 1623–1648.

Agrawal, A. (2005) *Environmentality: Technologies of government and political subjects*. Duke University Press, Duke.

Agrawal, A. and Angelsen, A. (2009) 'Using community forest management to achieve REDD+ goals' in A. Angelsen (ed.) *Realising REDD+: National strategy and policy options*, Center for International Forestry Research (CIFOR), Bogor, pp. 201–212.

Agrawal, A., Nepstad, D. and Chhatre, A. (2011) 'Reducing emissions from deforestation and forest degradation', *Annual Review of Environment and Resources*, vol 36, no 1, pp. 373–396. Available at: www.annualreviews.org/doi/10.1146/annurev-environ-042009-094508 (Accessed 30 January 2018).

Bardhan, P. (1984) *Political economy of India*. Oxford University, Delhi.

Baviskar, A. (1999) *In the belly of the river: Tribal conflicts over development in the Narmada Valley*. Oxford University Press, Delhi.

Baviskar, B. S. and Mathew, G. (2009) *Inclusion and exclusion in local governance: Field studies from rural India*. Sage Publications, Delhi.

Bayrak, M. M. and Marafa, L. M. (2016) 'Ten years of REDD+: A critical review of the impact of REDD+ on forest-dependent communities', *Sustainability*, vol 8, no 7, p. 620.

Bekele-Tesemma, A. and Ababa, A. (2002) *Forest landscape restoration: Initiatives in Ethiopia*. IUCN/WWF, Gland, Switzerland. Available from: http://assets.panda.org/downloads/ethiopiaflr.pdf (accessed 17 November 2009).

Beymer-Farris, B. A. and Bassett, T. J. (2012) The REDD menace: Resurgent protectionism in Tanzania's mangrove forests. *Global Environmental Change*, vol 22, no 2, pp. 332–341.

Bryant, R. L. and Bailey, S. (1997) *Third world political ecology*. Psychology Press, London and New York.

Büscher, B. (2012) 'Payments for ecosystem services as neoliberal conservation: (Reinterpreting) evidence from the Maloti-Drakensberg, South Africa', *Conservation and Society* vol 10, no 1, p. 29.

Carswell, G. and De Neve, G. (2013) 'Labouring for global markets: Conceptualising labour agency in global production networks', *Geoforum*, vol 44, pp. 62–70.

Chatterjee, P. (2008) 'Democracy and economic transformation in India', *Economic and Political Weekly*, pp. 53–62.

Chazdon, R. L., Brancalion, P. H., Laestadius, L., Bennett-Curry, A., Buckingham, K., Kumar, C. and Wilson, S. J. (2016) 'When is a forest a forest? Forest concepts and definitions in the era of forest and landscape restoration', *Ambio*, vol 45, no 5, pp. 538–550.

Chhatre, A. (2007) *Accountability in decentralization and the democratic context: Theory and evidence from India. Representation* (No. 23). Equity and Environment Working Paper.

Cornwall, A. (2007) 'Revisiting the "gender agenda"', *IDS Bulletin*, vol 38, no 2, pp. 69–78.

Cornwall, A. and Rivas, A. M. (2015) 'From "gender equality and women's empowerment" to global justice: Reclaiming a transformative agenda for gender and development', *Third World Quarterly*, vol 36, no 2, pp. 396–415.

Datta, B. (ed.). (1998) *And who will make the chapatis?: A study of all-women panchayats in Maharashtra*. Stree, distributed by Bhatkal Books International, USA.

Dudley, N., Mansourian, S. and Vallauri, D. (2005) 'Forest landscape restoration in context' in S. Mansourian, N. Dudley and D. Vallauri (eds) *Forest restoration in landscapes: Beyond planting trees*. Springer, Dordrecht.

FAO (2015) Forest Resource Assessment: Terms and definitions. *Forest Resources Assessment Working Paper 180*. Food and Agricultural Association of the United Nations, Rome.

Fairhead, J. and Leach, M. (1996) *Misreading the African landscape: Society and ecology in a forest-savanna mosaic* (Vol. 90). Cambridge University Press, Cambridge.

Fairhead, J., Leach, M. and Scoones, I. (2012) 'Green grabbing: A new appropriation of nature?', *Journal of Peasant Studies*, vol 39, no 2, pp. 237–261.

Fleischman, F. D. (2014) 'Why do foresters plant trees? Testing theories of bureaucratic decision-making in central India', *World Development*, vol 62, pp. 62–74.

Forsyth, T. (1996) 'Science, myth and knowledge: Testing Himalayan environmental degradation in Thailand', *Geoforum*, vol 27, no 3, pp. 375–392.

Fortmann, L. and Bruce, J. (1991) *You've got to know who controls the land and trees people use: Gender, tenure and the environment*. IDS, Sussex.

Franke, R. W. and Chasin, B. H. (1980) *Seeds of famine: Ecological destruction and the development dilemma in the West African Sahel*. Allanheld, Osmun and Co., Montclair, NJ.

Gupta, A. and Sivaramakrishnan, K. (eds). (2016) *The state in India after liberalization: Interdisciplinary perspectives*. Routledge, Canada.

Jaffrelot, C. (2003) *India's silent revolution: The rise of the lower castes in North India*. Orient Blackswan, Delhi.

Joshi, A. K., Pant, P., Kumar, P., Giriraj, A. and Joshi, P. K. (2011) 'National forest policy in India: Critique of targets and implementation', *Small Scale Forestry*, vol 10, no 1, pp. 83–96.

Karthik, M. and Kodiveri (2018) 'The great Indian land grab being carried out in the name of compensatory afforestation'. Available at: https://thewire.in/218841/great-indian-land-grab-carried-name-compensatory-afforestation/ (accessed 4 February 2018).

Kohli, K. and Menon, M. (2011) *Banking on forests: Assets for a climate cure?* Kalpavriksh, New Delhi.

Laestadius, L., Maginnis, S., Minnemeyer, S., Potapov, P., Saint-Laurent, C. and Sizer, N. (2011) 'Opportunities for forest landscape restoration', *Unasylva*, vol 62, no 2, p. 238.

Larson, A. M., Brockhaus, M., Sunderlin, W. D., Duchelle, A., Babon, A., Dokken, T., Pham, T. T., Resosudarmo, I. A. P., Selaya, G., Awono, A. and Huynh, T. B. (2013) 'Land tenure and REDD+: The good, the bad and the ugly', *Global Environmental Change*, vol 23, no 3, pp. 678–689.

Leach, M. (2007) 'Earth mother myths and other ecofeminist fables: How a strategic notion rose and fell', *Development and Change*, vol 38, no 1, pp. 67–85.

Lele, S. M. and Menon, A. (eds). (2014) *Democratizing forest governance in India*. Oxford University Press, Oxford.

Lund, J. F., Sungusia, E., Mabele, M. B. and Scheba, A. (2017) 'Promising change, delivering continuity: REDD+ as conservation fad', *World Development*, vol 89, pp. 124–139.

Martin, A. (2003) 'On knowing what trees to plant: Local and expert perspectives in the Western Ghats of Karnataka', *Geoforum*, vol 34, no 1, pp. 57–69.

McElwee, P. (2009) 'Reforesting "bare hills" in Vietnam: Social and environmental consequences of the 5 million hectare reforestation program', *Ambio*, vol 38, no 6, pp. 325–333.

McElwee, P. D. (2012) 'Payments for environmental services as neoliberal market-based forest conservation in Vietnam: Panacea or problem?', *Geoforum*, vol 43, no 3, pp. 412–426.

Meinzen-Dick, R. S., Brown, L. R., Feldstein, H. S. and Quisumbing, A. R. (1997) 'Gender, property rights, and natural resources', *World Development*, vol 25, no 8, pp. 1303–1315.

Mwangi, E., Meinzen-Dick, R. and Sun, Y. (2011) 'Gender and sustainable forest management in East Africa and Latin America', *Ecology and Society*, vol 16, no 1, article 17. Available at: www.ecologyandsociety.org/vol16/iss1/art17 (Accessed 19 June 2018).

Newton, A. C and Tejedor, N. (2011) 'Introduction' in Newton, A. C., ed., *Principles and Practice of Forest Landscape Restoration: Case Studies from the Drylands of Latin America*, IUCN, Gland, Chapter 1, pp. 1–22.

Norman, M. and Nakhooda, S. (2015) *The state of REDD+ finance*. Center for Global Development, Paper no 378.

Peskett, L. and Todd, K. (2012) 'Putting REDD+ safeguards and safeguard information systems into practice', *UNREDD Programme Policy Brief* Number 03.

Phelps, J., Webb, E. L. and Agrawal, A. (2010) 'Does REDD+ threaten to recentralize forest governance?', *Science*, vol 328, no 5976, pp. 312–313.

Poudel, M., Thwaites, R., Race, D. and Dahal, G. R. (2014) 'REDD+ and community forestry: Implications for local communities and forest management – a case study from Nepal', *International Forestry Review*, vol 16, no 1, pp. 39–54.

Rankin, K. N. (2001) 'Governing development: Neoliberalism, microcredit, and rational economic woman', *Economy and Society*, vol 30, no 1, pp. 18–37.

Razavi, S. (ed.). (2002) *Shifting burdens: Gender and agrarian change under neoliberalism*. Kumarian Press, Hartford, CT.

Resurreccion, B. P. (2006) 'Gender, identity and agency in Philippine upland development', *Development and Change*, vol 37, no 2, pp. 375–400.

Resurreccion, B. P. and Elmhirst, R. (2012) *Gender and natural resource management: Livelihoods, mobility and interventions*. Earthscan, London.

Reyes-García, V., Paneque-Gálvez, J., Bottazzi, P., Luz, A. C., Gueze, M., Macía, M. J., Orta-Martínez, M. and Pacheco, P. (2014) 'Indigenous land reconfiguration and fragmented institutions: A historical political ecology of Tsimane' lands (Bolivian Amazon)', *Journal of Rural Studies*, vol 34, pp. 282–291.

Robbins, P. (2011) *Political ecology: A critical introduction* (Vol. 16). John Wiley and Sons, London.

Robbins, P. and Tripuraneni, V. (2017, unpublished) *Colonial forest legacies: From Taux de Boisement to the national mission for green India*. Nelson Institute for Environmental Studies, University of Wisconsin, Madison.

Robertson, J. and Lawes, M. J. (2005) 'User perceptions of conservation and participatory management of iGxalingenwa forest, South Africa', *Environmental Conservation*, vol 32, no 1, pp. 64–75.

Sahu, G., Dash, T. and Dubey, S. (2017) 'Political economy of community forest rights', *Economic and Political Weekly*, vol 52, nos 25 and 26, pp. 44–47.

Sarin, M. (2005) *Laws, lore and logjams: Critical issues in Indian forest conservation*. Gatekeeper Series No. 116. International Institute for Environment and Development, London.

Scales, I. R. (2011) 'Farming at the forest frontier: Land use and landscape change in Western Madagascar, 1896–2005', *Environment and History*, vol 17, no 4, pp. 499–524.

Shiva, V. (1988) *Staying alive: Women, ecology and development*. Zed Books, New Delhi.

Skinner, E. (2011) *Gender and climate change. Overview report*. Institute of Development Studies, Sussex.

Sundar, N., Jeffery, R. and Thin, N. (2001) *Branching out: Joint forest management in India*. Oxford University Press, Delhi.

Sunderlin, W. D., Larson, A. M., Duchelle, A. E., Resosudarmo, I. A. P., Huynh, T. B., Awono, A. and Dokken, T. (2014) 'How are REDD+ proponents addressing tenure problems? Evidence from Brazil, Cameroon, Tanzania, Indonesia, and Vietnam', *World Development*, vol 55, pp. 37–52.

Svarstad, H. and Benjaminsen, T. A. (2017) 'Nothing succeeds like success narratives: A case of conservation and development in the time of REDD', *Journal of Eastern African Studies*, vol 11, no 3, pp. 482–505.

Tan, N. Q., Truong, L. T., Van, N. T. H., Enters, T., Yasmi, Y. and Vickers, B. (2010) *Evaluation and verification of the free, prior and informed consent process under the UN-REDD programme in Lam Dong Province, Vietnam*. The Center for People and Forests, Bangkok.

Temper, L. and Martinez-Alier, J. (2013) The god of the mountain and Godavarman: Net present value, indigenous territorial rights and sacredness in a bauxite mining conflict in India. *Ecological Economics*, vol 96, pp. 79–87.

Tiffen, M., Mortimore, M. and Gichuki, F. (1994) *More people, less erosion: Environmental recovery in Kenya*. John Wiley and Sons Ltd, London.

Van Noordwijk, M. and Minang, P. (2009) If we cannot define it, we cannot save it: Forest definitions and REDD. *ASB PolicyBrief*, 15.

Vira, B., Adams, B., Agarwal, C., Badiger, S., Hope, R. A., Krishnaswamny, J. and Kumar, C. (2012) 'Negotiating trade-offs: Choices about ecosystem services for poverty alleviation', *Economic and Political Weekly*, vol 47, no 9, pp. 67–76.

Williams, M. (2006) *Deforesting the Earth: From prehistory to global crisis, an abridgement*. University of Chicago Press, Chicago.

World Bank (2011) *Benefit sharing in REDD+: Policy note*. World Bank, Washington, DC. Available from: http://documents.worldbank.org/curated/en/205941 468340252680/Benefit-sharing-in-REDD-policy-note (Accessed 30 January 2018).

Part II

Approaches, systems and processes

5   Social-ecological systems and forest landscape restoration

Anastasia Yang, Imogen Bellwood-Howard and Melvin Lippe

Introduction

Forest landscape restoration (FLR) aims to regain ecological integrity and enhance human wellbeing, both important aspects of sustainability, requiring a sensitive mix of land uses to achieve the desired balance between the ecological and social elements of the system. Successful FLR entails understanding land use systems and the dynamics of land use change, for both planning and implementation. It also requires an understanding of how internal and external mechanisms interact with and influence resilience within those systems. It is argued that there is a need to achieve a more balanced focus in FLR efforts, particularly as it is common in restoration design to focus more on the ecological and less on the social aspects, despite the fact that these two systems are continuously interacting (Budiharta *et al.*, 2016). Land use decisions will be misplaced if based on overriding societal demands for immediate benefits over consequences for ecosystem functions. Similarly, to base FLR only on ecological criteria while ignoring the socio-political context would jeopardize the longevity and success of a scheme (Defries *et al.*, 2004; Budiharta *et al.*, 2016). This highlights the pressing need in FLR to understand how social components of a system interact with the ecological components via feedbacks, and how that influences thresholds (Walker and Meyers, 2004).

This chapter will explore how social-ecological systems (SES) frameworks provide a fitting conceptual approach to decision-making and can support FLR design, implementation and monitoring. SES frameworks can help establish a holistic understanding of both social and ecological interactions at nested scales, in a spatial and temporal context related to FLR practices. An SES approach can be used to examine whether external institutional and political settings, at multiple scales, are conducive to supporting FLR. Likewise, it can be used to help assess whether land users are liable to support or oppose forest restoration, depending on their attitudes and livelihood needs.

This chapter provides a justification for using SES to support FLR planning by outlining what SESs are, the key concepts around SES, with a focus on resilience, and how SES and FLR link, as well as the challenges of using SES for FLR. A range of forest and land management interventions, as methods for FLR, are outlined and analysed through an SES lens to identify how opportunities, trade-offs and ideas of resilience interact within ecological and social systems.

What are SESs and how do they link to FLR?

An SES is described as 'an ecological system intricately linked with and affected by one or more social systems' (Anderies *et al.*, 2004). An ecological system includes the sub-components of an interdependent group of organisms or biological resource units, such as trees, within a spatially explicit resource system such as a forest patch or national park (Anderies *et al.*, 2004; Ostrom, 2009; Bodin and Tengö, 2012). Ecosystem services are products of those systems, defined as 'the aspects of ecosystems utilized (actively or passively) to produce human well-being' (Fisher *et al.*, 2009). This definition itself underscores the acknowledged link between human welfare and functioning ecological systems. The social component focuses on the resource users and governance systems, from the scale of individuals, through communities and organizations, to nations and supranational entities (Bodin and Tengö, 2012). SES frameworks offer a means to analyse these combined ecological and social systems and recognize the interactions between them.

An SES approach can support FLR implementation, as it embraces interdisciplinary and multiscale analysis, aiming to understand the multi-faceted complexities of both ecological and social contexts (Yin and Zhao, 2012). Ostrom (2009) highlights that within specific scientific disciplines the language available to describe complex SESs can be limited (Cote and Nightingale, 2012); hence, there is a need to develop analytical approaches drawing on different disciplines to assist in integrating various disciplinary understandings. Research that stretches across multiple disciplines should emphasize the different sub-components of an SES not as separate, static, one-dimensional entities but as part of an interacting dynamic system, providing feedback mechanisms that influence subsystems of SESs at nested scales (Walker *et al.*, 2004; Janssen *et al.*, 2007; Ostrom, 2009) (see Figure 5.1, for example). SES theory is therefore promising for FLR, in that it does not shy away from complexity and embraces a holistic approach to research design and analysis. Cote and Nightingale (2012) argue that this can also open doors to a more inclusive style of resource and land management, incorporating lay observations and knowledge of ecological change. For example, Tengö and Hammer (2003) present a Tanzanian case study where farmers acknowledged the risk of environmental disturbances (e.g. floods and pest outbreaks) and, as a result, adopted more diverse land management practices to maintain a more stable agro-ecosystem.

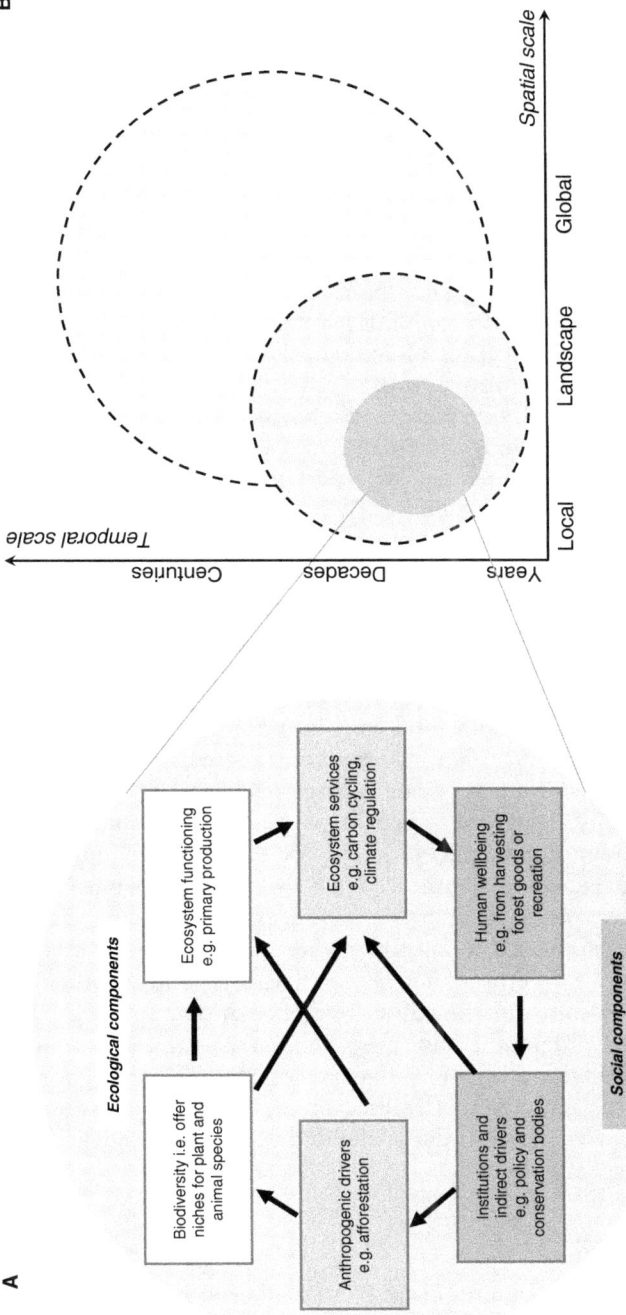

Figure 5.1 A. Example of an integrated FLR-SES in which the social components of institutions and policy and conservation bodies lead to a positive anthropogenic driver that fosters FLR. This, in turn, supports biodiversity, offers niches for plant and animal species, and improves ecosystem functioning, that is, primary productivity among other functions. Ecosystem services, which are the result of a healthy ecosystem, support carbon cycling and climate regulation on a regional and a global scale, and improve human wellbeing at local scale, for example by increasing the harvest of forest goods or providing recreation opportunities. B. Building on the SES framework, one still has to acknowledge that 5.1.A may be nested in a larger SES subsystem acting on a scale of decades (e.g. savannah woodlands), which is itself influenced by even larger temporal and spatial scales (e.g. ecoregion of sub-Saharan woodlands).

Source: adapted from Isbell *et al.*, 2017.

Key concepts in SES

SES embraces a range of concepts and terms. Understanding of the roles of multiple stakeholders, including land users, in addition to other actors who may have direct or indirect influence on and interest in land use decisions, is integral to understanding an SES. Two theoretical concepts further provide central notions about SES that are important in relation to its application to FLR. One is about multiscale spatial and temporal interactions; that is, effects across time, considering the past, present and future, and through space. This incorporates the notions of nested scales, panarchy and cascading effects. The second combines ideas related to resilience.

Buzz Holling's well-known (1973, 2001) version of SES recognizes different nested and interacting scales of ecological and social action. For example, a farm household is nested within a farming community, and each of these social structures is inextricably intertwined with its ecological environment. Processes happening at smaller scales, over shorter time periods, can influence those happening at larger, slower scales, and vice versa: this is termed 'cascading' (Holling, 2001). These processes involve cycles, in which a system moves from a more stable state, through periods of reorganization, to another more stable state.

The notion of cascading effects links to ideas of resilience and adaptive capacity. Resilience is described as the 'ability to withstand or maintain integrity in the face of a shock, and a switch to another ("steady") state' (Folke *et al.*, 2002). It is linked to adaptive capacity, the ability of the system to absorb changes, reorganize and shift into another state without significantly losing functionality. Adaptive capacity is more associated with human components of the SES: actors can manage resilience to improve a system's adaptive capacity (Walker *et al.*, 2004). In periods of stability, a system can remain below a threshold required for change, remaining in one regime and not attracted to another. When different disturbing and destabilizing forces rock it out of that system, and it is not resilient enough to absorb them, it may be pushed to another stable state instead. Its adaptive capacity can influence whether it moves to that state, and whether it persists in that state is also a function of its resilience and adaptive capacity. Switches between different states may happen at different scales (Walker and Meyers, 2004), and the non-linearity of SES further supports the idea that thresholds can move (DeFries *et al.*, 2004; Walker and Meyers, 2004). SES approaches consider social and ecological resilience as intertwined and interdependent.

Resilience of SES and FLR

Resilience is one of the key attributes of an SES (Walker *et al.*, 2004). SES holds that human action shapes ecosystem dynamics from local to global scales, while human societies rely on a wide variety of ecosystem services generated by SES for their wellbeing (Folke, 2006). In this view, change is an inherent characteristic of SES, not necessarily negative, but presenting ongoing opportunities for renewal and improvement (Holling, 2001; Walker *et al.*, 2006). A system's resilience determines how change within the system happens. In the framework of FLR, change relates to returning trees to land which may have been degraded or without forest cover. While restoration aims to build resilience of both people and nature, depending on choices made, resilience may not actually be strengthened. In particular, different tree species and forest communities will exhibit a different level of resilience under specific conditions.

Ecological resilience allows ecosystems to maintain themselves in the event of disturbance; for example, to maintain biodiversity or soil fertility during land use change. Social resilience can be defined as 'the ability of groups or communities to cope with external stresses and disturbances as a result of social, political, and environmental change' (Adger, 2000). There is a clear link between social and ecological resilience, particularly for social groups or communities that are dependent on ecological and environmental resources for their livelihoods. But it is not clear whether resilient ecosystems enable resilient communities in such situations (Adger, 2000). For instance, interconnectivity across scales has potential feedback loops, which can occur in bottom-up or top-down cascades.

The SES challenge for FLR

Understanding complexity and bridging interdisciplinary research

Although SES presents a useful guiding framework for FLR, the challenges should also be considered. For SES and FLR, interdisciplinary research is essential. Yet, bringing together the natural and social sciences is associated with practical and theoretical difficulties. At the research design stage, it can be challenging to develop a common team language and agree on the sample design and data collection and analytical approaches (Heberlein, 1988). Nevertheless, the advantages of interdisciplinary research are especially relevant to the joint project of SES and FLR, namely investigating complex interactions, feedbacks and holism between human and ecological components of a system; an SES approach in particular is about embracing complexity and seeking ways to understand it (Ostrom, 2009; Bodin and Tengö, 2012; Cotes and Nightingale, 2012). But how, then, can SES assist us to address the multifaceted issues associated with FLR and achieve a balance between ecological and social trade-offs and benefits?

An SES framework can provide the means to dissect complexity and combine the disciplines to understand different variables and how they are related. Binder *et al.* (2013) provide a useful comparison of established

SES frameworks used by researchers and practitioners, and criteria to determine which framework would be suitable for the type of issues that are being addressed. Frameworks can also be used to learn from and compare with other case studies, and to understand key characteristics that may determine relationships and outcomes in SESs (while acknowledging that unique characteristics will also be determinants).

System boundaries and reconciling landscape mosaics with nested scales (SES)

The reference scale of FLR is the landscape (Dudley *et al.*, 2005), perceived as a mosaic of dynamic patches, often at different stages of forest succession and regeneration (Chazdon, 2013), and themselves containing heterogeneous species mixes (Lamb *et al.*, 2012). The boundary and focus of FLR could be limited to that landscape; however, an SES approach would aim to broaden that view to also emphasize links within a panarchy of evolving hierarchical systems, with cascading effects. A landscape focus can also mask less positive heterogeneous outcomes at the local patch scale. For communities in a patch with high carbon sequestration potential yet few livelihood opportunities, landscape-scale multi-functionality is less relevant. Tensions and trade-offs thus emerge between different objectives at various scales, and the scale at which forest patches function ecologically does not necessarily match the scale at which they are managed (Cumming *et al.*, 2006). Sometimes, governance systems mould the shape of a restored landscape to fit social functioning, for example a national park, or community woodlots or sacred groves. The scale at which these interactions happen helps to construct the sustainability, resilience and stability of a landscape (Walker *et al.*, 2004).

Balancing ecological and social systems in FLR

What is, then, the optimal balance between ecological integrity, on the one hand, and people's livelihoods and wellbeing, on the other, considering the trade-offs between them that are inherent in FLR (DeFries *et al.*, 2004; Foley *et al.*, 2005)? FLR needs to consider dynamics across spatial and temporal scales, as well as heterogeneous landscapes, in order to build the optimal design, thereby acknowledging compromise between the ecological and social outcomes as inevitable.

There are a range of interventions to implement FLR that vary in terms of priorities, from those emphasizing restoration of ecosystem integrity to those that prioritize availability of socio-economic benefits. Figure 5.2 illustrates five forest and land management interventions that could be adopted to achieve FLR objectives, comparing how far each leads to ecological and social opportunities (5.2a and 5.2b), as well as how ecologically and socially resilient they are (5.2c and 5.2d).

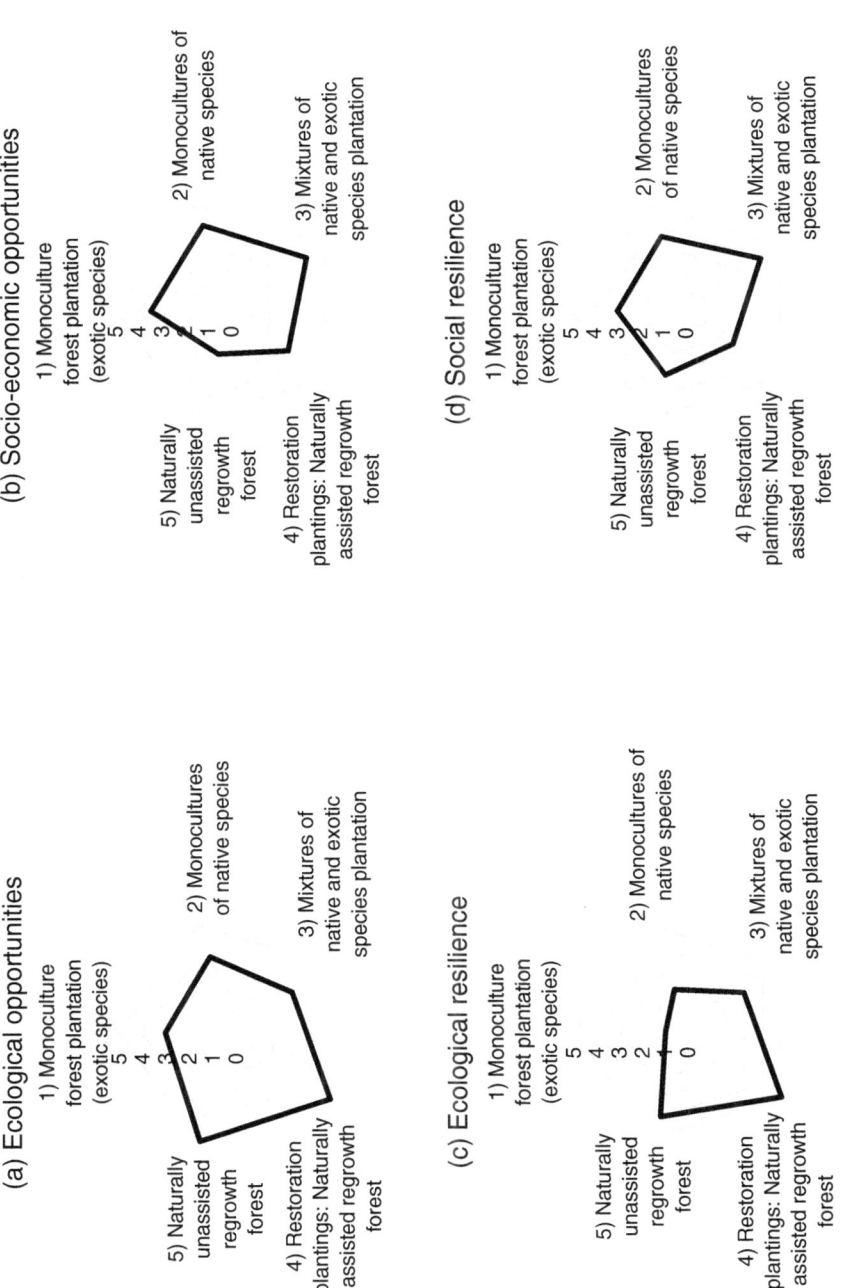

Figure 5.2 Five forest and land management interventions as methods to achieve FLR objectives and comparing SES parameters: social and ecological opportunities and resilience.

An overview of each of the five interventions emphasizes the differences between them (Lamb *et al.*, 2005):

1   *Exotic species monoculture forest plantation:* 'monoculture' refers to single species plantations, and 'exotic' refers to species that are introduced; often fast-growing, short-term tree species with commercial value are selected (e.g. eucalyptus, pine, rubber, acacia and teak).
2   *Native species monoculture forest plantation:* focuses on native tree species, mostly for productive purposes (e.g. fruits, timber fibre, fuelwood, charcoal and poles).
3   *Mixed plantation of native species (or with exotics):* often a mixture of tree and crop species, for instance agroforestry, focused on the production of diverse goods that can support a range of ecosystem services.
4   *Naturally assisted regrowth forest:* restoration plantings aim to restore ecosystem functionality and biodiversity via the planting of native species. Two main methods are: (i) using a small number of fast-growing, short-lived species to create canopy cover to facilitate colonization by a wider number of species, and (ii) using a higher number of species, avoiding the natural successional sequence and representing the mature successional stage.
5   *Naturally unassisted regrowth forest:* promotes restoring forest land to its pre-disturbance natural state with no assisted management. Suited to a more recently disturbed forest landscape.

Ecological and social opportunities

Ecological opportunity, as illustrated in Figure 5.2a, refers to the ability of a landscape or a land patch to support ecosystem functionality and provisional supporting and regulating services (Fisher *et al.*, 2009). The ecological opportunities offered by different sites vary, dependent on their past, present and potential future land uses. Social opportunities, as illustrated in Figure 5.2b, refer to the ability to derive both direct and indirect socioeconomic or livelihood benefits from the forest and land use interventions. These social opportunities reflect those hypothetically gained by mainly the local land users involved directly in delivering the intervention. Comparing the two figures, 5.2a and 5.2b, the trade-offs and synergies between the social and ecological opportunities of each intervention become apparent.

Monocultures offer ecological opportunity in cases where land is already highly degraded with poor soil conditions, long distances to seed sources, isolation, and invasion by aggressive grasses (Montagnini, 2005). Monoculture plantations can be driven by the offer of provisional ecosystem services for food and raw materials (e.g. timber) and/or the opportunity to promote soil and water conservation and rapid carbon sequestration (Chazdon, 2008). Yet carbon sequestration gains depend on the harvesting regime of the species: the longer rotations of native species plantations imply improved services. Equally, the ability of monoculture plantations to promote soil or water conservation is mixed, as they can also consume huge volumes of water (Scott, 2005). Ferraz *et al.* (2013) emphasize that this is largely dependent on forest management plans, with generally poorer water production and regulation services in fast-growing forest plantations. Furthermore, the emphasis of exotic monoculture tree plantations on the productivity of desired goods (i.e. timber) generally results in a direct trade-off with biodiversity gains, although these trade-offs can be minimized through appropriate plantation design and management actions (Brockerhoff *et al.*, 2008, 2013; Pawson *et al.*, 2013; Thompson *et al.*, 2014).

In terms of biodiversity, native monocultures could entail modest gains in comparison to exotics, but, as suggested by Bremer and Farley (2010), species richness gains are rare, and any gains will be largely dependent on the species introduced as well as the preceding land use characteristics, current land cover configuration across the landscape, and management practices (Montagnini, 2005; Brockerhoff *et al.*, 2008; Stanturf *et al.*, 2014). For instance, plantations are more favourable for ecological gains on degraded sites or those dominated by invasive plant species (Bremer and Farley, 2010). Yet such plantations, if they include invasive species, could risk the ecological integrity of surrounding areas, outcompeting other species, which could be to the detriment of adjacent forest areas. As indicated in Figure 5.2a, assisted and non-assisted natural forest regrowth have the potential to offer a larger number of ecological opportunities: in particular, the restoration of floral and faunal biodiversity, providing a wider array of ecosystem services such as pollination, air quality improvement, microclimate regulation and nutrient cycling (Lamb *et al.*, 2005; Hall *et al.*, 2012). In both cases, the speed and scale of regrowth depend strongly on soil and climate conditions, disturbance of stands, past land use and regional species pool (Chazdon and Guariguata, 2016).

A different pattern appears between the five land interventions when we look at social opportunities, as shown in Figure 5.2b, favouring the monoculture plantations. For example, the socio-economic 'commercial' gains in reforestation via natural regeneration are potentially more limited than those in monoculture plantations. Monocultures offer a moderate to high level of social opportunity as a comparatively efficient method for rapidly growing durable species. If highly degraded lands are targeted for forest plantations, surrounding or integrated agricultural production systems could also be improved, with benefits to soil quality and regulation of hydrology (Lusiana *et al.*, 2017), illustrating the link between ecological and social gains. The overall economic value of exotic timber will often be lower than for slow-growing native species, but yields can be higher. Alternatively, in native monoculture plantations, the long-term benefits can be greater, as market prices from slower-growing native timber species can be significantly higher and prices are increasing further as supplies from natural forests decline. Social opportunities with monoculture plantations are available at multiple scales, from large-scale industrial to smallholder levels.

The combination of crops and trees found in mixed planting systems, such as agroforestry, can balance the short- and long-term gains of perennials and annuals, as well as meeting subsistence and commercial needs. Under suitable management regimes, trees can also serve to moderate microclimatic conditions, improve soils and protect crops from pests, diseases and wind damage (Nair, 1993), adding resilience to these agroecological systems. On-farm trees also offer diverse sources of income to farmers: fruits, timber, fuelwood and medicines (Vira *et al.*, 2015).

For naturally assisted and non-assisted interventions, once forest systems become established, a range of non-timber forest products (NTFP) can be utilized by local people, which may also help to perpetuate and enhance traditional land management practices. Yet, there are potentially fewer immediate direct financial benefits to land owners, at least in the short term, due to slow natural regrowth, so these lands could be at risk of conversion to other more short-term profitable land use purposes.

Assisted natural regrowth aims to accelerate the establishment, growth and survival of native tree species through management (Shono *et al.*, 2007) in order to enrich degraded lands, but is limited by available planting material and seeds from natural forest in the vicinity of the sites. There is potential socio-economic benefit in natural unassisted regeneration, as it is the cheapest way of fostering reforestation over large areas, but only if forest species are still present on the landscape to initiate and sustain forest regrowth and the land conditions are still able to sustain those species within that system. Although unlikely to enable high yields of commercial products, it could, once established, offer alternative opportunities for ecotourism and recreation, especially if the enhanced biodiversity provides critical habitats for forest-dependent animal and plant species (Chazdon *et al.*, 2009; Poorter *et al.*, 2016).

Labour inputs for each of the interventions will also vary over time. Investment in assisted natural regrowth is intensive in the initial stages, but lower at later stages. Labour inputs for monoculture plantations can range in intensity but require inputs in the initial planting and maintenance, as well as harvesting (and potentially even processing and transporting of products), contingent on the management regime and set of species chosen. Unassisted natural regeneration inputs are minimal; hence, risks to livelihoods are also low, and labour would be focused on monitoring and protection.

Assisted and non-assisted natural forest regeneration systems can be more unpredictable, which may be unsatisfactory to those who wish to see an ordered restoration process with specific area-based and time-bound outcomes (Chazdon and Guariguata, 2016). Unassisted interventions could be viewed as unfavourable by governments, land owners and conservation organizations, as areas can be perceived as unmanaged. If land areas are perceived to be neglected or abandoned, they risk being converted by others for alternative purposes. For natural regrowth interventions to work, the appropriate socio-economic conditions and stakeholder support need to be matched to the suitable ecological conditions to allow defor-ested land to be converted back to natural forest, in most cases where opportunity costs for alternative land uses are low (Chazdon, 2008).

Ecological and social resilience

While resilience is linked to the opportunities an SES presents, the focus here is on their ability to resist and adapt to disturbances, avoiding undesirable cascade effects. Figure 5.2c illustrates the ecological resilience of the five interventions, and Figure 5.2d illustrates their potential social resilience. Of all the interventions, natural forest regrowth will have a rel-atively high disturbance threshold, and therefore ecological resilience, in part due to its heterogeneity. There may, however, be an extended period between disturbance and recovery. A further caveat is that if ecological conditions in targeted areas have surpassed a threshold and reached another steady state, for example due to topsoil loss or land cover change, restoring such areas to a natural state will be increasingly difficult and expensive (Lamb *et al.*, 2005).

Alternatively, moderate resilience could be achieved within a mixed planting system, such as agroforestry, which has a broader range of traits and functional performances, which may improve under changing environ-mental conditions (Matson *et al.*, 1997; Altieri, 1999; Lin, 2011). Resili-ence in monocultures is reduced because exotic species can be unsuited to local conditions and themselves may cause negative impacts on soils, water storage, stream flow and native biodiversity (Jackson, 2002; Farley *et al.*, 2005). Monoculture plantations of native species are also at high risk of disease and pest outbreaks, yet, compared with exotic species, are often better adapted to local environmental conditions; and seeds could be more readily available.

For social resilience, Figure 5.2d illustrates that the ability of the five interventions to offer stability to local land users' livelihoods ranges from moderate to high, largely dependent again on the restoration location in question. As with ecological systems, social resilience depends on previous and current conditions at nested scales. Nevertheless, exotic and native monoculture plantations imply moderate livelihood resilience, as both entail substantial investment of money and time, while cultivation of a single species implies a vulnerability to social or ecological distur-bances. Native species could offer more stability to such disturbance, and high-value tree species can serve as a 'safety-net' for farmers, to pay for schooling or medical fees as well as food security in times of scarcity. Equally, the amount of commercially viable and available forest products influences how far a system can contribute socially. Thus, the diverse range of species present in mixed and natural systems could lead to higher livelihood resilience, for example in the event of market or climatic fluctuations.

The knowledge level of land managers is also fundamental to the social resilience of each intervention, largely due to their ability to mitigate and adapt to disturbances, whether natural or anthropogenic. As the complexity of the plantation system increases, so does the necessary knowledge on ecological and silvicultural practices; for example, how to collect seeds and germinate seedlings for multiple species plantation. Thus, native species may take longer to reach maturity, but farmers are likely to be more familiar with their use and management (Montagnini, 2005). Technical knowledge and skills are needed in the case of mixed planting systems and naturally assisted forest regrowth interventions, and livelihood resilience could be strengthened if these areas are well managed and workers' skill base improved. For exotic species, land managers may initially have limited knowledge on their suitability, but this could be addressed with adequate technical assistance and access to guidelines.

This overview highlights various pertinent issues, such as the implications of maintaining balances between fast-growing and slow-growing as well as indigenous and exotic species; the ability to derive economic and ecological benefits and restore multiple ecosystem functions; and the speed at which each of these may happen. Each of the five examples of forest and land management interventions has different implications for the resilience of both the ecological and the social components of a forest system, depending on the conditions and levels of resource dependency of local users. The distribution of benefits among users may also be unequal. The livelihood opportunities benefit mostly local land users, as these actors will be most likely directly affected by and/or involved in FLR. Yet this may not always be the case, as areas under consideration for FLR will also include a range of tenure regimes and owners, including foreign private land holders, and/or the government. Nevertheless, the provision of commercial forest goods is an important benefit that acts across spatial scales. Where commercial goods are limited, the success of an intervention could depend on the availability of alternative livelihood options and the level of local forest dependency. Natural regrowth interventions focus strongly on the restoration of ecosystem functions and returning forests to a natural state, which has broad benefits beyond the local level. At the local scale, the amount of forest products that could be commercially viable may be prioritized, but not assumed, again depending on local land users' experience and understanding of SES feedback mechanisms.

In addition, ecosystem benefits gained from the range of FLR approaches are enhanced as approaches move away from monoculture plantations to more species-diverse restoration practices. The proportion of gains to trade-offs will depend on the scale of implementation of specific restoration actions. Therefore, mosaic landscape approaches are more suited than are site-specific initiatives to achieve a compromise (Lamb *et al.*, 2005). The suitability of one approach over another is largely dependent on the current ecological status of the land under consideration. Once an ecosystem has passed a certain threshold, the ability to restore its ecological integrity will be severely compromised. Thus, a step-wise approach may be required, in the awareness that a fuller range of SES benefits will only be realized gradually over time, via a series of compromises. For example, in Denmark, it took 200 years to achieve more multifunctional forests, starting initially with monoculture plantations of Norway spruce to reclaim severely degraded soils (Madsen *et al.*, 2005). The nested governance system should also be considered, involving household and community-level decision-making capacities, but also thinking about how the local and national governance systems could either facilitate or impede FLR success. In general, what is determined as the optimal compromise will depend on the context and on the needs and interests of all stakeholders.

Conclusion

The aim of FLR is to reconcile both ecological and social objectives to restore forests at the landscape level: this may not always be possible at the patch level. An SES approach leads to the question of whether there are always trade-offs, and whether trade-offs can be minimized at the landscape scale. Does the landscape approach gloss over inequalities that people and other living species may experience at the local scale? How do these potential trade-offs affect system resilience? These are questions that can be addressed through an SES lens. Lessons learned can be drawn from FLR approaches that have already been established, and from similar forest-related initiatives such as community-based forest management schemes and payments for ecosystem services, which have also been implemented at the landscape scale.

So what is the outlook for FLR, from an SES perspective? The definition of what to 'restore' will be case specific, and a point well reflected in the FLR literature (Lamb *et al.*, 2005). The value of SES in supporting FLR lies partly in its ability to frame and structure complex issues to support decisions on what, where and how FLR approaches should be implemented. An SES approach can help to interrogate the focal scale of an FLR effort and recognize the potential strengths and weaknesses of certain approaches for supporting ecological and social systems. A balance between ecological and social resilience seems possible in certain situations if the landscape mosaic includes a combination of restoration approaches, with patch-specific actions linked to a wider landscape effort. This can link back to the notion of nested scales in SES, as explored earlier in 'Key concepts in SES', where subsystems can contribute to the functioning of the overall system. Yet, this does not imply making local-scale outcomes subservient to those at the landscape scale, and especially here, SES should serve as a tool to remember this and assess its implications in specific contexts.

Both SES and FLR examine how to balance the resilience of human and ecological components of forest systems. They force a consideration of how, in the event of deforestation, the integrity of a landscape system would persist, in terms of factors such as slope stability and soil fertility as well as community cohesion and human health. FLR needs the kind of interdisciplinary, integrated approach advocated by SES. Thus, baseline interdisciplinary research can help to understand systems at multiple spatial and temporal scales, taking account of contextual needs and interests. However, the impetus and support for forest landscape restoration will need to be generated at the local level as much as it will be needed by government actors.

Trade-offs often exist between resilience in the human and ecological components of a system, where some system regimes are desirable by one segment of society while undesirable by another (Walker *et al.*, 2006). Therefore, what is more important from an FLR perspective is, rather, the question of: 'Resilience of what and for whom?' (Cote and Nightingale, 2012). Accordingly, resilience cannot be measured in absolute terms. In a further example, many tropical forest landscape ecosystems can have stable and diverse populations but relatively low resilience, whereas temperate landscapes with apparently low diversity can exhibit greater resilience (Adger, 2000). An optimal win-win situation will be challenging to achieve in this context. Yet, if managers can strive to enhance and link the resilience of both the social and the ecological components of a system, it will support the overall effectiveness of FLR.

In summary, SES is equally relevant as a management approach as it is a research tool in the case of FLR. It can lead to an emphasis on multi-stakeholder engagement and decision-making. It can also encourage landscape managers to seek connections beyond forestry as a sector, interacting with other sectors such as agriculture, fisheries, recreation and manufacturing at a landscape scale. Overall, as Ostrom *et al.* (2007) highlight, while no solution is a panacea, using a diagnostic approach such as that developed for SES could increase the prospects for future sustainable forest resource use.

References

Adger, W.N. (2000) 'Social and ecological resilience: Are they related?', *Progress in Human Geography*, vol 24, no 3, pp. 347–364.
Altieri, M.A. (1999) 'The ecological role of biodiversity in agroecosystems', *Agriculture, Ecosystems & Environment*, vol 74, no 1, pp. 19–31.
Anderies, J., Janssen, M. and Ostrom, E. (2004) 'A framework to analyze the robustness of social-ecological systems from an institutional perspective', *Ecology and Society*, vol 9, no 1, article 18. Available from: www.ecologyandsociety.org/vol9/iss1/art18/ (accessed 19 June 2018).

Binder, C.R., Hinkel, J., Bots, P.W. and Pahl-Wostl, C. (2013) 'Comparison of frameworks for analyzing social-ecological systems', *Ecology and Society*, vol 18, no 4, article 26. Available from: http://dx.doi.org/10.5751/ES-05551-180426 (accessed 19 June 2018).

Bodin, Ö. and Tengö, M. (2012) 'Disentangling intangible social–ecological systems', *Global Environmental Change*, vol 22, no 2, pp. 430–439.

Bremer, L.L. and Farley, K.A. (2010) 'Does plantation forestry restore biodiversity or create green deserts? A synthesis of the effects of land-use transitions on plant species richness', *Biodiversity and Conservation*, vol 19, no 14, pp. 3893–3915.

Brockerhoff, E.G., Jactel, H., Parrotta, J.A., Quine, C.P. and Sayer, J. (2008) 'Plantation forests and biodiversity: Oxymoron or opportunity?', *Biodiversity and Conservation*, vol 17, no 5, pp. 925–951.

Brockerhoff, E.G., Jactel, H., Parrotta, J.A. and Ferraz, S.F. (2013) 'Role of eucalypt and other planted forests in biodiversity conservation and the provision of biodiversity-related ecosystem services', *Forest Ecology and Management*, vol 301, pp. 43–50.

Budiharta, S., Meijaard, E., Wells, J.A., Abram, N.K. and Wilson, K.A. (2016) 'Enhancing feasibility: Incorporating a socio-ecological systems framework into restoration planning', *Environmental Science & Policy*, vol 64, pp. 83–92.

Chazdon, R.L. (2008) 'Beyond deforestation: Restoring forests and ecosystem services on degraded land', *Science*, vol 320, pp. 1458–1460.

Chazdon, R.L. (2013) 'Making tropical succession and landscape reforestation successful', *Journal of Sustainable Forestry*, vol 32, no 7, pp. 649–658.

Chazdon, R.L. and Guariguata, M.R. (2016) 'Natural regeneration as a tool for large-scale forest restoration in the tropics: Prospects and challenges', *Biotropica*, vol 48, no 6, pp. 716–730.

Chazdon, R.L., Peres, C.A., Dent, D., Sheil, D., Lugo, A.E., Lamb, D., Stork, N.E. and Miller, S.E. (2009) 'The potential for species conservation in tropical secondary forests', *Conservation Biology*, vol 23, no 6, pp. 1406–1417.

Cote, M. and Nightingale, A.J. (2012) 'Resilience thinking meets social theory: Situating social change in socio-ecological systems (SES) research', *Progress in Human Geography*, vol 36, no 4, pp. 475–489.

Cumming, G., Cumming, D.H. and Redman, C. (2006) 'Scale mismatches in social-ecological systems: Causes, consequences, and solutions', *Ecology and Society*, vol 11, no 1, article 14. Available from: www.ecologyandsociety.org/vol11/iss1/art14/ (accessed 19 June 2018).

DeFries, R.S., Foley, J.A. and Asner, G.P. (2004) 'Land-use choices: Balancing human needs and ecosystem function', *Frontiers in Ecology and the Environment*, vol 2, no 5, pp. 249–257.

Dudley, N., Morrison, J., Aronson, J. and Mansourian, S. (2005) 'Why do we need to consider restoration in a landscape context?' pp. 51–58 in S. Mansourian, D. Vallauri and N. Dudley (eds), *Forest Restoration in Landscapes, Beyond Planting Trees*, Springer Netherlands, Dordrecht.

Farley, K.A., Jobbágy, E.G. and Jackson, R.B. (2005) 'Effects of afforestation on water yield: A global synthesis with implications for policy', *Global Change Biology*, vol 11, no 10, pp. 1565–1576.

Ferraz, S.F., de Paula Lima, W. and Rodrigues, C.B. (2013) 'Managing forest plantation landscapes for water conservation', *Forest Ecology and Management*, vol 301, pp. 58–66.

Fisher, B., Turner, R.K. and Morling, P. (2009) 'Defining and classifying ecosystem services for decision making', *Ecological Economics*, vol 68, no 3, pp. 643–653.

Foley, J.A., DeFries, R., Asner, G.P., Barford, C., Bonan, G., Carpenter, S.R., Chapin, F.S., Coe, M.T., Daily, G.C., Gibbs, H.K. and Helkowski, J.H. (2005) 'Global consequences of land use', *Science*, vol 309, no 5734, pp. 570–574.

Folke, C. (2006) 'Resilience: The emergence of a perspective for social–ecological systems analyses', *Global Environmental Change*, vol 16, no 3, pp. 253–267.

Folke, C., Carpenter, S., Elmqvist, T., Gunderson, L., Holling, C.S. and Walker, B. (2002) 'Resilience and sustainable development: Building adaptive capacity in a world of transformations', *AMBIO: A Journal of the Human Environment*, vol 31, no 5, pp. 437–440.

Hall, J.M., Van Holt, T., Daniels, A.E., Balthazar, V. and Lambin, E.F. (2012) 'Trade-offs between tree cover, carbon storage and floristic biodiversity in reforesting landscapes', *Landscape Ecology*, vol 27, no 8, pp. 1135–1147.

Heberlein, T.A. (1988) 'Improving interdisciplinary research: Integrating the social and natural sciences', *Society & Natural Resources*, vol 1, no 1, pp. 5–16.

Holling, C.S. (1973) 'Resilience and stability of ecological systems', *Annual Review of Ecology and Systematics*, vol 4, no 1, pp. 1–23.

Holling, C.S. (2001) 'Understanding the complexity of economic, ecological, and social systems', *Ecosystems*, vol 4, no 5, pp. 390–405.

Jackson, W. (2002) 'Natural systems agriculture: A truly radical alternative', *Agriculture, Ecosystems & Environment*, vol 88, no 2, pp. 111–117.

Janssen, M.A., Anderies, J.M. and Ostrom, E. (2007) 'Robustness of social-ecological systems to spatial and temporal variability', *Society and Natural Resources*, vol 20, no 4, pp. 307–322.

Lamb, D., Erskine, P.D. and Parrotta, J.A. (2005) 'Restoration of degraded tropical forest landscapes', *Science*, vol 310, no 5754, pp. 1628–1632.

Lamb, D., Stanturf, J. and Madsen, P. (2012) 'What is forest landscape restoration?' pp. 3–23 in J. Stanturf, D. Lamb and P. Madsen (eds) *Forest Landscape Restoration: Integrating Natural and Social Sciences*, World Forests 15, Springer Netherlands, Dordrecht.

Lin, B.B. (2011) 'Resilience in agriculture through crop diversification: Adaptive management for environmental change', *BioScience*, vol 61, no 3, pp. 183–193.

Lusiana, B., Kuyah, S., Öborn, I. and van Noordwijk, M. (2017) 'Typology and metrics of ecosystem services and functions as the basis for payments, rewards and co-investment' in S. Namirembe, B. Leimona, M. van Noordwijk and P.A. Minang (eds) *Co-investment in Ecosystem Services: Global Lessons from Payment and Incentive Schemes*, World Agroforestry Centre, Nairobi.

Madsen, P., Jensen, F.A. and Fodgaard, S. (2005) 'Afforestation in Denmark' in J. Stanturf and P. Madsen (eds) *Restoration of Boreal and Temperate Forests*, CRC Press, Boca Raton.

Matson, P.A., Parton, W.J., Power, A.G. and Swift, M.J. (1997) 'Agricultural intensification and ecosystem properties', *Science*, vol 277, no 5325, pp. 504–509.

Montagnini, F. (2005) 'Selecting tree species for plantation. Forest Restoration in Landscapes', pp. 262–268 in S. Mansourian, D. Vallauri and N. Dudley (eds) *Forest Restoration in Landscapes, Beyond Planting Trees*, Springer Netherlands, Dordrecht.

Nair, P.R. (1993) *An Introduction to Agroforestry*, Kluwer Academic Publishers Group, Dordrecht.

Ostrom, E. (2009) 'A general framework for analyzing sustainability of social-ecological systems', *Science*, vol 325, no 5939, pp. 419–422.

Ostrom, E., Janssen, M.A. and Anderies, J.M. (2007) 'Going beyond panaceas', *Proceedings of the National Academy of Sciences*, vol 104, no 39, pp. 15176–15178.

Pawson, S.M., Brin, A., Brockerhoff, E.G., Lamb, D., Payn, T.W., Paquette, A. and Parrotta, J.A. (2013) 'Plantation forests, climate change and biodiversity', *Biodiversity and Conservation*, vol 22, no 5, pp. 1203–1227.

Poorter, L., Bongers, F., Aide, T.M., Zambrano, A.M.A., Balvanera, P., Becknell, J.M., Boukili, V., Brancalion, P.H., Broadbent, E.N., Chazdon, R.L., Craven, D., de Almeida-Cortez, J.S., Cabral, G.A., de Jong, B.H., Denslow, J.S., Dent, D.H., DeWalt, S.J., Dupuy, J.M., Durán, S.M., Espírito-Santo, M.M., Fandino, M.C., César, R.G., Hall, J.S., Hernandez-Stefanoni, J.L., Jakovac, C.C., Junqueira, A.B., Kennard, D., Letcher, S.G., Licona, J.C., Lohbeck, M., Marín-Spiotta, E., Martínez-Ramos, M., Massoca, P., Meave, J.A., Mesquita, R., Mora, F., Muñoz, R., Muscarella, R., Nunes, Y.R., Ochoa-Gaona, S., de Oliveira, A.A., Orihuela-Belmonte, E., Peña-Claros, M., Pérez-García, E.A., Piotto, D., Powers, J.S., Rodríguez-Velázquez, J., Romero-Pérez, I.E., Ruíz, J., Saldarriaga, J.G., Sanchez-Azofeifa, A., Schwartz, N.B., Steininger, M.K., Swenson, N.G., Toledo, M., Uriarte, M., van Breugel, M., van der Wal, H., Veloso, M.D., Vester, H.F., Vicentini, A., Vieira, I.C., Bentos, T.V., Williamson, G.B. and Rozendaal, D.M. (2016) 'Biomass resilience of neotropical secondary forests', *Nature*, vol 530, no 7589, pp. 211–214.

Scott, D.F. (2005) 'On the hydrology of industrial timber plantation', *Hydrological Processes*, vol 19, pp. 4203–4206.

Shono, K., Cadaweng, E.A. and Durst, P.B. (2007) 'Application of assisted natural regeneration to restore degraded tropical forestlands', *Restoration Ecology*, vol 15, no 4, pp. 620–626.

Stanturf, J.A., Palik, B.J. and Dumroese, R.K. (2014) 'Contemporary forest restoration: A review emphasizing function', *Forest Ecology and Management*, vol 331, pp. 292–323.

Tengö, M. and Hammer, M. (2003) *Management Practices for Building Adaptive Capacity: A Case from Northern Tanzania*, Cambridge University Press, Cambridge.

Thompson, I.D., Okabe, K., Parrotta, J.A., Brockerhoff, E., Jactel, H., Forrester, D.I. and Taki, H. (2014) 'Biodiversity and ecosystem services: Lessons from nature to improve management of planted forests for REDD-plus', *Biodiversity and Conservation*, vol 23, no 10, pp. 2613–2635.

Vira, B., Wildburger, C. and Mansourian, S. (eds) (2015) *Forests and Food: Addressing Hunger and Nutrition across Sustainable Landscapes*, Open Book Publishers, Cambridge.

Walker, B. and Meyers, J.A. (2004) 'Thresholds in ecological and social–ecological systems: a developing database', *Ecology and Society*, vol 9, no 2, issue 3. www.ecologyandsociety.org/vol9/iss2/art3/.

Walker, B., Holling, C.S., Carpenter, S. and Kinzig, A. (2004) 'Resilience, adaptability and transformability in social–ecological systems', *Ecology and Society*, vol 9, no 2. www.ecologyandsociety.org/vol9/iss2/art5/.

Walker, B.H., Gunderson, L.H., Kinzig, A.P., Folke, C., Carpenter, S.R. and Schultz, L. (2006) 'A handful of heuristics and some propositions for understanding resilience in social-ecological systems', *Ecology and Society*, vol 11, no 1, article

13. Available at: www.ecologyandsociety.org/vol11/iss1/art13/ (accessed 19 June 2018).

Yin, R. and Zhao, M. (2012) 'Ecological restoration programs and payments for ecosystem services as integrated biophysical and socioeconomic processes – China's experience as an example', *Ecological Economics*, vol 73, pp. 56–65.

6 Integrated landscape approaches to forest restoration

Jeffrey Sayer and Agni Klintuni Boedhihartono

Introduction

The landscapes that we observe today, throughout the world, are the product of continuous change over time. The objective of forest landscape restoration (FLR) should never be to lock the landscape into a configuration that exists at a single point in time, nor should it be to restore the landscape to a condition that existed at some arbitrary point in the past. Perhaps this has been the weakness of many attempts to restore or rehabilitate landscapes – they have been driven by a single vision of what the ideal landscape might be. The reality is that people's needs and aspirations will continue to change and there will be no permanent consensus on an ideal landscape. Society has to guard against the concept of 'designer landscapes' conceived by 'experts' and representing the views of a subset of stakeholders. Governance processes must address this need for constant negotiation and adaptation to meet constantly changing needs and possibilities. A particular problem has been that forest restoration projects are usually driven by forest departments or interest groups who seek to maximize forest area and timber production. Forest institutions may be blind to the interests of other users of the land and may pursue narrow sectoral views of what constitutes an optimal landscape. Social media campaigns following the launch of the Bonn Challenge on FLR with its targets to restore millions of hectares of forested landscapes were a reaction to the perception that FLR might represent an attempt at 'land grabbing' by forest departments and companies. FLR might pose a threat to the livelihoods of other users of those landscapes. This chapter presents experience of ways of approaching FLR without the constraints of sectoral or disciplinary interests. We explore a number of techniques that enable the landscape to be considered in a holistic way to ensure that the legitimate interests of all stakeholders are addressed in an equitable and balanced way.

Landscape degradation

While degradation is essentially a subjective condition (e.g. Hobbs, 2016), any initiative to restore forests in landscapes must attempt to understand what drivers of change have led to the present condition of the landscape. What are the root causes of forest and landscape degradation? We contend that degraded landscapes are almost always a symptom of bad landscape governance. Regulations and decision-making processes have failed to address the competing claims of different landscape actors, with the result that landscape values of importance to some interest groups have been lost at the expense of other interest groups. Degradation often results from powerful interest groups appropriating land or favouring short-term appropriation of private benefits at the expense of long-term husbandry of public benefits. The need for FLR comes from the absence of effective rules that would prevent some people or companies from degrading the landscapes at the expense of other people or companies. As the world's population inexorably increases and demands on resources intensify, there is increased competition for natural resources, and degradation is often the result. The importance of appropriate and effective governance arrangements to achieve a balance between the interests of different stakeholders is therefore becoming greater. The stakeholders may be local communities in search of land to cultivate crops but may also include the global community concerned with biodiversity preservation or with carbon stocks to mitigate climate change. Competition may be for immediate access to timber, wildlife or land, or it may be between people concerned with the immediate wellbeing of their families and others aiming for long-term global sustainability. The need for FLR is often the result of the absence of effective rules to prevent some individuals from profiting at the expense of other individuals. Differentials of power are often underlying causes of landscape degradation. For example, vast swathes of forest landscapes in the tropics that are classified as state land are also claimed by local communities under customary tenure; resulting conflicts frequently lead to the clearance of forest as a way to stake a claim (e.g. Sunderlin *et al.*, 2009; Mansourian *et al.*, 2016).

Any initiative to restore forests in a landscape must be based upon a holistic and integrated vision of how the landscape might look in the future. It has to be based upon a level of agreement among stakeholders, and, again, it must not be constrained by the narrow interests of the forest sector or determined by the specific tools of forest disciplines. An interdisciplinary approach is essential and must underpin any major FLR initiative. We use the term 'landscape approach' to describe the set of tools and approaches that we have used to balance social and ecological concerns in large heterogeneous geographical spaces. The terms 'jurisdictional approach' and 'area-based approach' are essentially similar to 'ecosystem' and 'integrated rural development' approaches, which have been used in the past. Concepts and tools have evolved, but all these terms describe attempts that take an inclusive and integrated approach to managing landscape change (Sayer and Campbell, 2004; Sayer *et al.*, 2007).

Landscape approach and forest landscape restoration

We define a landscape approach as 'a long-term collaborative process bringing together diverse stakeholders aiming to achieve a balance between multiple and sometimes conflicting objectives in a landscape or seascape' (Sayer *et al.*, 2016). The basic principles that make up a landscape approach are described in Sayer *et al.* (2013). A landscape approach is, therefore, quite different from a technology-led planning approach whereby experts determine the optimal composition or configuration of a landscape. A landscape approach will rarely have an endpoint and will not usually produce a definitive 'plan' for the future of the landscape. It will more usually be a continuing process of negotiation and adaptation to ensure that an optimal balance is met between the conflicting goals of different stakeholders. Landscape approaches are underpinned by technical tools that can aid in these processes of negotiation (Sayer *et al.*, 2013), but ultimately it is the actors in the landscape who will determine by their decisions and behaviours the way in which the landscape changes over time. The knowledge and interests of local actors will be at least as important as those of external interest groups, and 'citizen science' will often be more relevant that outside expertise (Sayer *et al.*, 2015b).

FLR will always be problematic. No single arrangement of forests will be ideal for all actors, and the needs of actors will change over time. Landscapes are the product of multiple decisions made by diverse actors over long periods of time. The pattern of woodlands and hedgerows in Europe has evolved from decisions made about land enclosure many hundreds of years ago, and these patterns will continue to evolve as the social and economic environment of the actors continues to change. The iconic irrigated rice landscapes of Bali in Indonesia and Ifegau in the Philippines are the result of continuous processes of adaptation and change that have endured for many centuries, and these also will continue to evolve in response to changing opportunities and constraints imposed by the modern cash economy. A landscape is, therefore, a complex and continually evolving system, and understanding of its dynamics will need to draw upon expertise from different interest groups and different sectors. Landscapes are theatres where interdisciplinary science has to be deployed to achieve understanding, and any interventions to change or 'restore' landscapes must, by the same token, be driven by integrative science (Campbell and Sayer, 2003). Difficulties arise when interventions to change landscapes are driven by people who have a narrow disciplinary or sectoral approach – they may fail to recognize attributes of the landscape that are of significance to people from other disciplines or sectors.

Landscape governance

We define governance as the arrangements that societies put in place to take decisions and to ensure that these decisions are enforced. Local and regional governments tend to be the bodies that bring together and represent actors in a landscape. Landscapes may or may not be subject to formal systems of governance. Yet, governments should ensure that decisions about landscapes are inclusive of the interests of all legitimate stakeholders. FLR is widely needed because governmental bodies have failed to achieve such a balance between competing claims. Landscape problems may often occur at the level of jurisdictions, and the bodies that govern these jurisdictions will need to lead efforts to achieve restoration. The reality, however, is that often restoration efforts are led by specialized agencies that have a mandate to protect watersheds, conserve habitat of rare species, or extend forest cover that has been degraded by misuse or by uncontrolled exploitation. Landscapes often do not align with formal jurisdictions; in fact, the reasons why a landscape is degrading may lie in the fact that its values cross the boundaries of jurisdictions.

A first step in embarking upon an FLR initiative will often be to understand the system of governance that is in place. Surveys will be required to determine who is taking and enforcing decisions and to assess the measures that are in place to achieve coherence among the decision-makers. In most cases, it will be found that networks of actors are influencing a landscape, and often these networks are surprisingly complex. In any FLR initiative, it is of fundamental importance to understand these governance arrangements. Simply interviewing people and institutions that have agency in the landscape is a first step. Ensuring that all legitimate parties have been consulted and that their views are being considered is essential. Often, people and institutions will have conflicting or competing claims on land and its resources. Understanding the nature of the conflicts and competition is important. People and institutions that have 'agency' – that is, who have the capacity to influence landscape-level outcomes – can be identified and the nature of their interests noted. The degree to which their interests are aligned or are in competition can be tabulated. Simple box and arrow diagrams showing the interactions between these actors can be useful in showing relationships and clarifying points of conflict. Software tools such as Mindmap can be used to give a formal representation of institutional relationships. The different actors who have agency – whose combined actions and decisions provide the governance of the system – must be parties to multi-stakeholder discussions. In an ideal situation, a stakeholder platform will be the key governance body for a landscape (Boedhihartono, 2012).

Serious efforts to understand and influence complex governance arrangements will require the use of actor network theory and can be facilitated by using specialized software that can display the relationships among actors. For example, we used Gephi software to analyse actor networks in landscapes where restoration is taking place on the Indonesian island of Lombok (see Figure 6.1).

This research showed that the assumptions made by Indonesian government agencies and outside aid agencies about the way governance was operating in the landscape were not always correct. Government officers particularly assumed that governance was hierarchical and operated in a linear fashion within sectors. The reality was that there were numerous overlapping and often conflicting agencies making decisions about the landscape, with results that were somewhat contradictory and inefficient.

The term 'polycentric governance' is used to describe situations where governance is distributed among different agencies and across different geographical (and jurisdictional) scales (Ostrom, 2010; Chapter 11, this volume). Polycentric governance can be effective and efficient if coordination and cooperation exist among the different decision-making bodies, but it can also be ineffective and result in problematic outcomes if

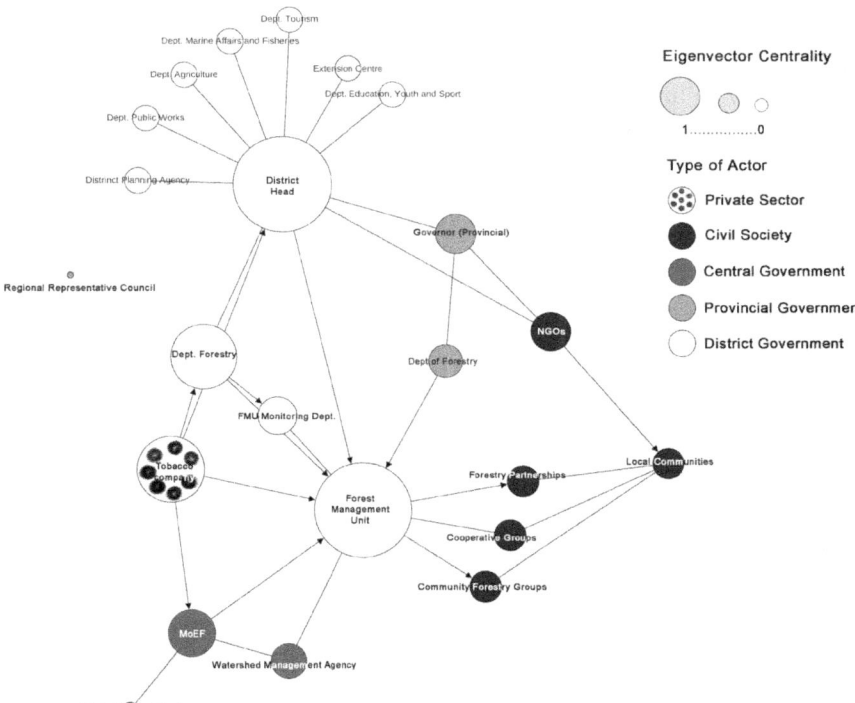

Figure 6.1 Actor network diagram generated using Gephi software showing the complexity of the relationships among institutions that influence the governance of a landscape restoration initiative on the Indonesian island of Lombok.

Source: adapted from Riggs et al. (2018).

cooperation and collaboration are not effective. If some agencies simply have more power than others, then the results of polycentric governance will often be unsatisfactory. Forest departments have been notable for not paying enough attention to the actions of other interest groups and not recognizing the trade-offs between forests and other competing land uses. Models that clarify decision-making can be helpful in taking decisions on how to intervene to achieve landscape outcomes.

Actor network studies will reveal who needs to be involved in order to improve decision-making processes. If a landscape is to be restored, then the behaviour of many agencies – people and institutions – will need to be influenced. This is where stakeholder forums or platforms are needed. Stakeholder forums allow representatives of interest groups to gather to discuss and negotiate landscape change. Convening and facilitating stakeholder forums is itself a complex and often challenging task. Stakeholders will have conflicting views, and it may be difficult to reach agreement among them. Balint *et al.* (2011) documented the difficulties of managing multi-stakeholder forums and used the term 'wicked problems' to describe the difficulties encountered by the US Forest Service in maintaining the engagement of stakeholders and reaching agreement on how to restore landscapes. Balint ultimately concluded that the Forest Service had to listen to all actors but then take the best and most equitable decisions on the basis of a professional assessment of possible actions. Multi-stakeholder forums will rarely be able to resolve entrenched disagreements and also will rarely have the authority to impose their decisions. They can contribute to better governance, but they can often only forge a degree of consensus and often lack the authority to truly 'govern' a landscape. Landscapes should, in principle, be governed by jurisdictional bodies – local planning boards and other local government agencies – and landscape approaches are often used when these formal government bodies are not producing satisfactory outcomes. Landscape approaches are a response to failure of formal jurisdictional governance mechanisms.

Achieving an understanding of governance arrangements and an ability to intervene to improve governance failings is a fundamental first step in any attempt at integrated landscape restoration. Many reforestation programmes of forest departments have been controversial because governance arrangements were not effective or were not considered legitimate by stakeholders, who felt that their views were not being taken into account. Intergovernmental meetings of the Bonn Challenge, which is seeking support for ambitious FLR goals, have led to demonstrations and social media criticism. Governments and their forest agencies were accused of land grabbing – seeking to expand the forest areas under their jurisdiction at the expense of other land users. It is fundamental to any landscape restoration effort to answer the question of 'restoring what and for whom' (Boedhihartono and Sayer, 2012). Even if there is wide agreement that the proportion of the landscape under forest should increase, there will remain the questions of what sort of forest, whom it will belong to, and who will have the right to use it. Landscape restoration programmes, for example, in China, Vietnam, Indonesia and the Mediterranean countries of Europe have all encountered problems because they proceeded without proper consultation on the type of forest that stakeholders wanted. Forest departments have often tended to plant those species that they were most familiar with or that met an immediate market demand. These species may not have been optimal for the diversity of uses that local people required from the forest.

Tools and methods

Understanding and documenting historical change: A valuable tool in gaining an integrated and holistic understanding of landscape dynamics is to engage local stakeholders in an analysis of past events that have influenced the landscape (see Table 6.1). In its simplest form, this can be achieved by asking people to mark on a sheet of paper or a white board the events that they recall that were significant in shaping the landscape. Such exercises often lead to animated discussions about the landscape, the landscape features that people value, and how they have gained or lost from changes. Developing historical timelines can be a useful first step in gaining stakeholder engagement in discussions of how a landscape might be improved and, hence, of what FLR should target. Historical timelines will often reveal that the present landscape has been shaped by numerous external and internal forces over many decades or even longer.

Such discussions lead to increased understanding of the complexity and impacts of external influences and internal behaviours. The production of a historical timeline will often lead seamlessly into a discussion of what the future of the landscape might be and what the worst-case and best-case scenarios for the future might include. FLR will be just one of many initiatives that will shape the future landscape, and understanding how it will interact with other drivers of change is important. The discussions around historical change will enrich and enable discussions of desirable and undesirable futures for the landscape.

Landscape scenarios – visualizing desired future landscapes: Perhaps the single most powerful technique for gaining an integrated understanding of how landscape function impacts on the lives of different stakeholders is the use of visualization approaches (see Figure 6.2). At their simplest, these may consist of asking groups of stakeholders to sketch on a paper or white board their understanding of the present and potential future condition of the landscape. They should be encouraged to portray in their drawings the features of the landscape that they consider to be important to their livelihoods. It is often effective to ask these groups to draw best-case and worst-case scenarios for landscapes. The actors in the landscape can be divided into interest groups – women and men may have different views on what

Table 6.1 A good way of engaging local stakeholders in a discussion about changes in their landscape is to ask them to list all the historical events that have influenced the landscape. This historical profile shows the major events that have shaped the landscape of the island of Boano in the Indonesian Province of Maluku

Year	Events
2000–1500 BCE	Chinese traders visit Maluku Islands
1460–1465 CE	Islam arrives in the region
1511	Portuguese arrive in Luhu on the mainland opposite Boano
1599	Dutch arrive in Maluku Wooden boats being built on Boano
1726	Visit of Dutch trader Valentinj and naming of Teluk Valentinj site of settlement on Boano
1950s	Making houses using coral and sand Use of fences around *kebun* (gardens) to protect from wild boar
1959	SR (*Sekolah Rakyat*) – elementary school started
1960s	Wooden houses constructed
1965	Start to have Indonesian military (TNI) members recruited from and based on Boano Private junior high school established
1970s	Coffee cultivation expanded
1977	SD (*Sekolah Dasar*) – public primary school replaced the *Sekolah Rakyat*
1979	Starvation on Boano after drought
1980s	Cacao cultivation expands
1980–1981	Boat linking Boano with Ambon starts (two boats operating: *Taman Murni* and *Mawana Indah*)
1985	SMP (*Sekolah Menengah Pertama*) – public junior high school started on Boano
1986	Chinese trader starts processing *Minyak Kayu Putih* (*Melaleuca* oil) on Boano
1987	Logging on Boano – PT Busana Indah concession starts exploiting, looking for Kayu Lassi (Intsia sp.)
1988	Community expands collecting *Kayu Putih* for therapeutic oil production
Late 1980s	A gastro-intestinal disease, *muntaber*, spreads on the island
1990s	Start of building houses using *batako* (concrete blocks) First students from Boano enrol at University in Ambon *Puskesmas* – dispensary/health clinic opens in South Boano
1993/1994	Boat service from Boano to Seram – from and to *Pelita* and *Masika*

Table 6.1 Continued

Year	Events
1997	Long dry season (El Nino) – nine months – causing a drought
1999	Inter-communal conflict in Ambon and Seram People cease cultivating dry hill rice (*beras merah*) Before 1999, only five students from Boano graduated with a Bachelor's degree
2000s	*Puskesmas* – health clinic started in Boano North
2001–2002	Solar panels introduced
2004–2005	Electricity provided to main settlements on Boano island More Boano people going to University
2005	Fresh water pipeline installed but broke after several months (still not fixed)
2007–2014	*PNPM* (government grants to community) funded activities on Boano
2009	Three Boano people become members of DPRD (elected assembly for the Province)
2010s	*Sekolah Menengah Atas* – primary school started on Boano
2010	Fibreglass boats start to be used on Boano
2012	The long bridge across a flooded area on Boano is made
2014	Only one member of DPRD from Boano Water piping installed again (but only works for a few months)
2014–2015	Floods in the village, and houses are surrounded by water
2015	Agriculture programme from *Dinas Pertanian* (Ministry of Agriculture) – agricultural crops (seedlings for lemoncina, nutmeg, clove) and cows
2017	*Dana Desa* (village support grant from government) used for getting kettles (stills) for making *Kayu Putih* – 120 kettles in 2017 and 60 more kettles in 2018
2018	Plan for running water (pipe system) into houses

would constitute a desirable landscape, and younger people and older people might also differ in their views. Boedhihartono (2012) describes the techniques that can be used to exploit the potential of visualization. The different representations by different categories of actors can be made permanent by laminating them with plastic and then kept as guides to future interventions and also as a record of what was decided. Laminated scenario drawings can be valuable for monitoring and evaluation of landscape initiatives such as FLR.

When alternative visions for the future of the landscape are being debated, actors can be encouraged to vote for their preferred choice; they

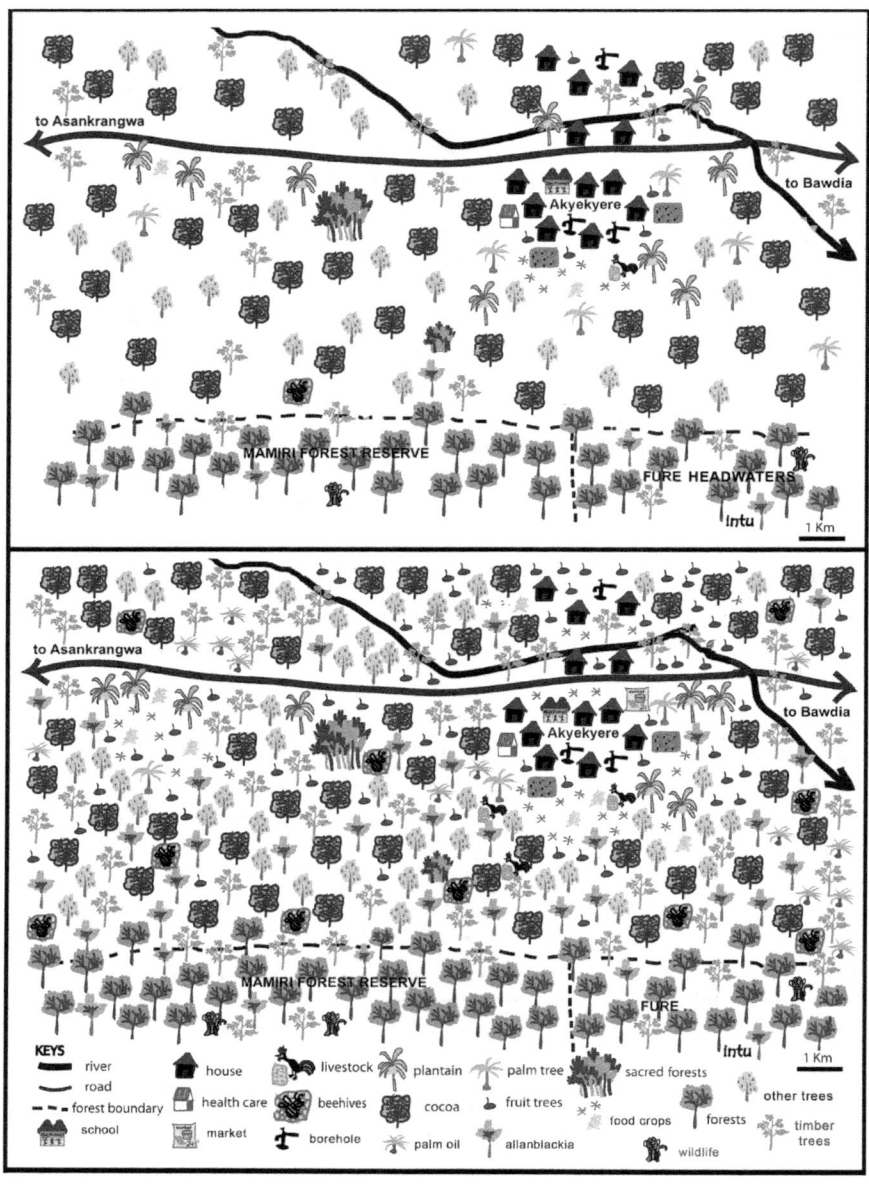

Figure 6.2 Drawings produced by local stakeholders can be influential in enriching the debate about desired future landscapes. These drawings were made by people from Akyekyere village in Western Ghana and show their wish to have a greater diversity of tree crops in their preferred future landscape.

can be given a number of sticky dots and allowed to post them onto the landscapes and landscape features that they approve of or disapprove of – we have described this process as 'dotmocracy' (Boedhihartono, 2012). Visual techniques used with good facilitation can break down disciplinary and sectoral barriers, lead to a more holistic understanding of the landscape, and contribute to building consensus on the desirable future that FLR can provide (Boedhihartono and Sayer, 2012).

Predicting impacts of FLR actions: Landscapes are complex systems. It is notoriously difficult to predict the impacts of interventions and to know what the future holds in store. Distant events may have more impact on a landscape than local interventions such as tree planting. Sayer *et al.* (2012) found that the global financial crisis had greater impact on forest conservation in a Congo Basin forest landscape than the local interventions of conservation non-governmental organizations (NGOs) and aid agencies. Sandker *et al.* (2007) used models to show that although oil palm expansion in East Kalimantan, Indonesia, was seen as a threat to forest conservation by many NGOs, there were scenarios where oil palm expansion would reduce pressure for shifting cultivation on remaining forest areas and would therefore be good for forest outcomes. Simulation models can be valuable tools for exploring the complexity of landscapes and determining who will be the losers and who the beneficiaries of different approaches to landscape restoration (Collier *et al.*, 2011). Models may be developed by experts and used to explore the implications of different future scenarios, but they can also be used with the participation of local people to help them to fully understand the risks and benefits of landscape change (Sandker *et al.*, 2010). Wu and Hobbs (2002) have argued that landscapes are inherently so complex that they can only be fully understood by using models, and many landscape ecologists now rely on simulation modelling to share understanding of landscape dynamics. Any major FLR initiative should consider investing in the development of a simulation model of its target landscape. Models allow the modeller to answer 'what if' questions and to apply scientific rigour in decision-making about land use trade-offs. The impact of tree planting on agricultural livelihoods or on hydrological functions can be modelled, and models can be used as 'decision support' tools in FLR planning (van Noordwijk *et al.*, 2001).

The role of citizen science: The inhabitants of a landscape will inevitably have greater knowledge of the values of a landscape than many outside experts. Change will impact on the lives of these people, and they will be aware of and concerned about any change that is planned or imposed. They will often make better judgements on the likely impacts of any FLR initiatives than outside planners. Stakeholder forums and historical analysis will provide ways of accessing the knowledge of people, and 'citizen science' can be a valuable approach to engage people in experimenting and learning about landscape change. The involvement of citizens can lead to their engagement and support for FLR. Exploiting citizen science can ensure that the FLR processes make sense for these local actors. Ensuring that citizen knowledge is mobilized to support restoration is something that needs to be explicitly pursued in any landscape initiative (Sayer *et al.*, 2015b).

Mediating landscape change: The significant investments that are being proposed for FLR and the importance of the environmental and social benefits that FLR is expected to underline the importance of ensuring effective decision-making processes. The leaders of FLR initiatives must have legitimacy with concerned stakeholders. They must ensure that all stakeholders are able to have their views heard. Negotiations must be equitable and transparent – the powerful must not be allowed to dominate the less powerful. Feedback must be allowed – those impacted by FLR activities must be allowed to comment on progress and report back on any negative or positive impacts that they experience. There must be a continuous process of learning and adaptation. It is relatively easy to convene multi-stakeholder forums, but it is challenging to ensure that they function effectively. When decisions are made, institutions must be in place and capable of enforcing agreements.

The different tools and approaches outlined in this chapter must be deployed in the correct sequence; feedback and adaptation must be effective (Boedhihartono and Sayer, 2012). The question of who leads FLR initiatives must be addressed – and there is no simple answer to this question. Outside agencies – aid agencies and international NGOs – often seek to provide leadership, but they may not be seen to have legitimacy by local actors. Government agencies should provide leadership, but often they bear some responsibility for the degradation that has occurred, and they may not be ideal leaders of restoration. Government agencies often align with narrow sectoral interests and may not be seen as 'neutral' facilitators. Someone will have to initiate the process, and this will usually entail convening a multi-stakeholder gathering, but the subsequent leadership of that gathering should ideally be determined by the stakeholders themselves. A successful multi-stakeholder process in the Congo Basin was facilitated by an international NGO (Endamana *et al.*, 2010).

Measuring success

Throughout this chapter, we have emphasized that FLR outcomes cannot be defined precisely in advance; even if one may aim for a vision of the restored landscape, there will inevitably be continuous negotiations and adjustments to the outcomes that are being sought. Measuring progress towards intermediate objectives will help to drive adaptive management strategies. However, there is no definitive endpoint – no point at which the landscape can be declared to have been restored. Some objectives may have been reached, such as reducing fragmentation or improving pollination, but their permanence will require long-term management. The reality of continuing uncertainty means that conventional approaches to measuring impact or success may not be applicable. Most assessments of the performance of FLR initiatives will need to be based upon process indicators and indicators of short-term outputs or outcomes. Long-term impacts at the level of improvements in social or ecological conditions will only be identified and measurable over time. The difficulties of measuring success and some of the approaches to the measurement of process are discussed in full in Sayer *et al.* (2016).

Preconditions for effective forest landscape restoration

The likelihood that FLR initiatives will succeed is highly dependent on context. Several preconditions have been identified for landscape approaches that greatly increase the chances of success (Sayer *et al.*, 2015a). We have attempted to adapt these and apply them to FLR, and we suggest that the following are important.

1 **Inspired leadership** is essential: Leadership may come from formal institutions, but frequently it is provided by committed individuals. Champions such as the International Union for Conservation of Nature (IUCN) and the Global Partnership on Forest and Landscape Restoration (GPFLR) can empower FLR initiatives and give them international exposure.
2 **Long-term, adaptive commitment:** FLR is a long-term process, and any initiative must receive long-term but flexible support. All too often, FLR (and other restoration) projects are driven by donors functioning on short-term cycles, which do not reflect the long time frames necessary for both restoration and social adaptation.
3 **Facilitation** is necessary but not sufficient to negotiate trade-offs when considering where to restore forests within a landscape.
4 **Value propositions** will motivate engagement: People will engage with landscape-scale processes and multi-stakeholder forums if there is sufficient reason for them to do so. Restoration, in particular, can shift the value of a landscape – with different consequences, including creating new beneficiaries. Restoration can also provoke conflicts and create power differentials, with new interest groups gaining prominence.
5 **Conflict and entrenched views must be openly addressed:** Facilitation alone will not reconcile fundamentally conflicting interests.
6 **Strong systemic governance is essential:** Agreements related to FLR have to be enforceable by law, cadastral records need to be in place, institutions need to be functioning, and land rights need to be clear and accepted.
7 **Private sector engagement** is a key element of success. Given the scale of the FLR challenge, funding from the private sector is increasingly being sought, as public funding will not suffice.

8 **Policies without budgets and implementation commitments do not work**: The institutions responsible for implementing FLR have to be adequately funded and have the mandate to enforce agreements. However, examples exist in South America of 'overregulation' related to restoration (and forests more generally), and this has limited their effectiveness because of resistance from local stakeholders.

9 **Formalization and monitoring of process outcomes are eventually needed**: Agreements may be reached and activities influenced, but eventually, FLR must be integrated into legal measures to ensure sustainable long-term outcomes.

10 **Metrics must be developed to establish values, track progress and enable adaptive management**: Societies' demands upon landscapes will change over time, and adaptive management will be needed. Metrics must be put in place to enable the flows of goods and services from the landscape to be monitored, and programmes will need to be changed if demands are not being met.

Conclusions

FLR, like other landscape interventions, is not a precise science. It will always be an inherently messy process, and it will confront dissenting views and 'wicked problems'. The reason why forest restoration is now widely advocated at a landscape scale is the recognition that the values of forests in a landscape have an impact on many divergent stakeholders. We conduct restoration at a landscape scale precisely because we recognize the need to take into account the interests of diverse interest groups. FLR must be underpinned by integrative science. The concepts of 'sustainability science' – science inspired by the needs of the users of the science – need to be applied to FLR (Clark, 2007). No two FLR processes will be identical, and there can be no 'cookie-cutter' approaches to FLR. Every situation is unique, and leading these processes will be like jazz – a constant process of learning and adaptation. However, there are basic principles that can be drawn upon to guide the process of landscape approaches, and these have been described by Sayer *et al.* (2013). The following guidelines are an adaptation of these principles for FLR.

- **Continuous learning and adaptation:** It is important not to be locked into a rigid set of activities and to retain the ability to learn and adapt throughout the FLR process.
- **Common concern/problem:** Stakeholders will engage in FLR activities if it makes sense for them personally to do so – FLR will work if the participants share a common concern or problem.
- **Multiple scales:** FLR is distinguished from other restoration activities because it takes into account impacts at different spatial and temporal

scales. The impacts of restoration measures on both distant and local stakeholders have to be reconciled.

- **Multi-functionality:** The 'restored' landscape will have to provide a balance between the needs of divergent stakeholders. The landscape may provide timber, support biodiversity and protect hydrological functions. FLR, by definition, must provide multiple functions.
- **Multiple stakeholders:** Landscapes provide values to many different stakeholders, and it is essential that the needs and wishes of all stakeholders are taken into account.
- **Theory of change:** Restoring a landscape will require changes in the behaviour of institutions and individuals. Careful thought has to be given to the measures that will influence this behaviour. A theory of change must be designed by stakeholders and must provide for feedback and adaptation as FLR progresses.
- **Clear rights and responsibilities:** FLR will impact upon the livelihoods of stakeholders. Success will be influenced by the behaviour of these stakeholders, and their behaviour will be more likely to contribute to broader landscape goals if their rights and responsibilities are clear and enforced.
- **Participatory monitoring:** The ultimate beneficiaries of FLR should be local stakeholders, and these stakeholders should lead the process of assessing the success or failure of FLR. A prime role of a multi-stakeholder forum is to provide feedback to the process. The beneficiaries of FLR must lead the process of assessing impact.
- **Resilience:** FLR must take into account the reality that the forces for landscape change will not remain unchanged over time. Landscapes will be impacted by external forces such as climate change, population movements, economic adjustments and so on. FLR must provide for and contribute to the resilience of the landscape. This may require some built-in redundancy in the form of extra checks and balances, but it especially requires that landscape stakeholders see their assets improving over time. Assets include those of **individuals** – education and health; of **societies** – institutions, laws and social networks; **nature** – soils, forests and biodiversity; **physical infrastructure** – roads, water supplies, buildings; and **finance** – income, profits, savings.
- **Capacity:** FLR will need capacity; this will lie in the capability of individuals and institutions to implement all of the interventions, processes and governance arrangements discussed in this chapter.

FLR is a reaction against traditional sectoral and discipline-driven approaches to natural resource management. It exemplifies the need for natural resource management to be integrated if it is to address the complex problems facing the world. FLR should be at the cutting edge of interdisciplinary science and be rooted in a holistic understanding of the problems of sustainable development. FLR is where 'sustainability science' hits the road. Most importantly, the implementation of FLR will require the deployment of skilled professionals able to operate across disciplinary boundaries and to empathize with the interests of multiple and diverse stakeholders.

References

Balint, P. J., Stewart, R. E., Desai, A. and Walters, L. C. (2011) *Wicked environmental problems: Managing uncertainty and conflict*, Island Press, Washington.

Boedhihartono, A. K. (2012) *Visualizing sustainable landscapes: Understanding and negotiating conservation and development trade-offs using visual techniques*, IUCN, Gland.

Boedhihartono, A. K. and Sayer, J. (2012) 'Forest landscape restoration: Restoring what and for whom?' in Stanturf, J., Lamb, D. and Madsen, P. eds. *Forest landscape restoration: Integrating natural and social sciences* (Vol. 15). Springer, Dordrecht.

Campbell, B. M. and Sayer, J. A. (2003) *Integrated natural resource management: Linking productivity, the environment and development*, CABI Publishing, Wallingford.

Clark, W. C. (2007) 'Sustainability science: A room of its own', *Proceedings of the National Academy of Sciences of the United States of America*, vol 104, pp. 1737–1738.

Collier, N., Campbell, B. M., Sandker, M., Garnett, S. T., Sayer, J. and Boedhihartono, A. K. (2011) 'Science for action: The use of scoping models in conservation and development', *Environmental Science and Policy*, vol 14, pp. 628–638.

Endamana, D., Boedhihartono, A., Bokoto, B., Defo, L., Eyebe, A., Ndikumagenge, C., Nzooh, Z., Ruiz-Perez, M. and Sayer, J. A. (2010) 'A framework for assessing conservation and development in a Congo Basin Forest Landscape', *Tropical Conservation Science*, vol 3, pp. 262–281.

Hobbs, R. J. (2016) 'Degraded or just different? Perceptions and value judgements in restoration decisions', *Restoration Ecology*, vol 24, no 2, pp. 153–158.

Mansourian, S., Razafimahatratra, A., Ranjatson, P. and Rambeloarisoa, G. (2016) 'Novel governance for forest landscape restoration in Fandriana-Marolambo, Madagascar', *World Development Perspectives*, vol 3, pp. 28–31.

Ostrom, E. (2010) 'Beyond markets and states: Polycentric governance of complex economic systems', *American Economic Review*, vol 100, pp. 1–33.

Riggs, R.A., Langston, J.D., Margules, C., Boedhihartono, A.K., Lim, H.S., Sari, D.A. Sururi, Y. and Sayer, J. (2018) 'Governance Challenges in an Eastern Indonesian Forest Landscape', *Sustainability*, vol 10, no 1, pp. 169–187. doi:10.3390/su10010169.

Sandker, M., Campbell, B. M., Ruiz-Perez, M., Sayer, J. A., Cowling, R., Kassa, H. and Knight, A. T. (2010) 'The role of participatory modeling in landscape approaches to reconcile conservation and development', *Ecology and Society*, vol 15, no. 2, article 13. Available from: www.ecologyandsociety.org/vol15/iss2/art13/ (accessed 19 June 2018).

Sandker, M., Suwarno, A. and Campbell, B. M. (2007) 'Will forests remain in the face of oil palm expansion? Simulating change in Malinau, Indonesia', *Ecology and Society*, vol 12, no. 2, article 37. Available from: www.ecologyandsociety. org/vol12/iss2/art37 (accessed 19 June 2018).

Sayer, J. and Campbell, B. M. (2004) *The science of sustainable development: Local livelihoods and the global environment*, Cambridge University Press, Cambridge.

Sayer, J., Margules, C., Boedhihartono, A. K., Dale, A., Sunderland, T., Supriatna, J. and Saryanthi, R. (2015a) 'Landscape approaches: What are the pre-conditions for success?', *Sustainability Science*, vol 10, no 1, pp. 345–355.

Sayer, J., Margules, C., Bohnet, I., Boedhihartono, A., Pierce, R., Dale, A. and Andrews, K. (2015b) 'The role of citizen science in landscape and seascape approaches to integrating conservation and development', *Land*, vol 4, pp. 1200–1212.

Sayer, J., Sunderland, T., Ghazoul, J., Pfund, J.-L., Sheil, D., Meijaard, E., Venter, M., Boedhihartono, A. K., Day, M. and Garcia, C. (2013) 'Ten principles for a landscape approach to reconciling agriculture, conservation, and other competing land uses', *Proceedings of the National Academy of Sciences of the United States of America*, vol 110, pp. 8349–8356.

Sayer, J. A., Endamana, D., Ruiz-Perez, M., Boedhihartono, A. K., Nzooh, Z., Eyebe, A., Awono, A. and Usongo, L. (2012) 'Global financial crisis impacts forest conservation in Cameroon', *International Forestry Review*, vol 14, pp. 90–98.

Sayer, J. A., Maginnis, S. and Laurie, M. (2007) *Forests in landscapes: Ecosystem approaches to sustainability*, Earthscan, London.

Sayer, J. A., Margules, C., Boedhihartono, A. K., Sunderland, T., Langston, J. D., Reed, J., Riggs, R., Buck, L. E., Campbell, B. M. and Kusters, K. (2016) 'Measuring the effectiveness of landscape approaches to conservation and development', *Sustainability Science*, vol 12, no 2, pp. 465–476.

Sunderlin, W. D., Larson, A. M. and Cronkleton, P. (2009) 'Forest tenure rights and REDD' in A. Angelsen (ed.) *Realising REDD*, Center for International Forestry Research (CIFOR), Bogor, pp. 139–150.

van Noordwijk, M., Tomich, T. P. and Verbist, B. (2001) 'Negotiation support models for integrated natural resource management in tropical forest margins', *Conservation Ecology*, vol 5, no 2, p. 21.

Wu, J. and Hobbs, R. (2002) 'Key issues and research priorities in landscape ecology: An idiosyncratic synthesis', *Landscape Ecology*, vol 17, pp. 355–365.

7 Forest landscape restoration and land sparing–sharing

Shifting the focus towards nature's contributions to people

Agnieszka E. Latawiec,
Juliana Silveira dos Santos, Veronica Maioli,
André B. Junqueira, Renato Crouzeilles,
Catarina C. Jakovac, Fernanda Tubenchlak
and Bernardo B. N. Strassburg

Introduction

Isolating agricultural production from natural areas is no longer possible in many countries where high-diversity forest ecosystems are currently embedded in human-modified landscapes (Scherr and McNeely, 2008; Janishevski *et al.*, 2015). As natural forest ecosystems and agricultural landscapes provide essential services for human wellbeing, there is an indisputable need for strategies that reconcile food production with biodiversity conservation.

Forest landscape restoration (FLR) aims at regaining ecological integrity and enhancing human wellbeing as well as enhancing landscape functionality in deforested or degraded forest landscapes (Mansourian *et al.*, 2005; Maginnis and Jackson, 2007). A fundamental feature of FLR is the combination of forest and non-forest ecosystems balanced with different land uses, also known as the 'landscape approach' (Sayer *et al.*, 2013; see also Chapter 6, this volume). The landscape approach focuses on allocating and managing land to achieve environmental and socio-economic benefits in landscapes where multiple land uses coexist and where environmental goals often compete with agriculture and other land uses (Sayer, 2009; Sayer *et al.*, 2013). This approach is also relevant to the United Nations Sustainable Development Goals (SDGs), which call for 'holistic and integrated approaches to sustainable development that restore the health and integrity of the Earth's ecosystem'. At least five of the key objectives of the SDGs overlap with the landscape approach: (i) to end hunger; (ii) to secure water; (iii) to promote strong, inclusive and sustainable economic growth; (iv) to tackle climate change; and (v) to protect and promote terrestrial resources.

A variety of restoration strategies can be implemented within the landscape approach in the search for solutions to reconcile trade-offs between biodiversity conservation, agricultural production and human development. How to achieve the best balance of land uses, biodiversity conservation and ecosystem services provision is a complex question, closely related to the land sparing and land sharing debate. Both approaches support the fundamental argument that the integration of food production and nature conservation is possible, as will be discussed in the following section.

More recently, a new way to move forward the land sparing and sharing debate has been proposed by broadening the focus to human wellbeing, ecosystem services (Bennett, 2017) and 'Nature's Contributions to People' (NCP) (IPBES, 2016; Pascual *et al.*, 2017). Under this view, valuing landscapes only for their biodiversity conservation or food production potential ignores other important services. Agricultural landscapes are inherently multifunctional, providing a range of ecosystem services such as aesthetics, recreation, and water and carbon regulation (Bennett, 2017). An integrated view requires, therefore, that wider benefits from landscapes at different scales are accounted for rather than aiming to maximize the two main goals of either land sparing or sharing: food production and biodiversity conservation (Bennett, 2017).

While ecosystem services approaches have been present in the literature for much longer, NCP is a very recent concept proposed by the Intergovernmental Platform on Biodiversity and Ecosystem Services (IPBES; Pascual *et al.*, 2017). The term 'Nature's Contributions to People' broadens the ecosystem services approach as proposed by the Millennium Ecosystem Assessment (MEA, 2005) and is defined by 'all the positive contributions or benefits, and occasionally negative contributions, losses or detriments that people obtain from nature'. It resonates with the use of the term 'ecosystem services' and goes further by explicitly embracing concepts associated with other worldviews on human–nature relations and knowledge systems (e.g. 'nature's gifts' in many indigenous cultures) (Pascual *et al.*, 2017). The authors claim that understanding the diversity of values of NCP requires recognizing a broad range of worldviews on nature and respecting various constituents that translate to a good quality of life.

Promoting different conceptualizations of value and valuation, therefore, requires engaging in bridging and mobilizing transdisciplinary collaboration across natural and social sciences, and other knowledge systems (see Chapter 12, this volume). Furthermore, decision-making processes must be value-laden to achieve societal goals for sustainability (Pascual *et al.*, 2017). While ecosystem services have been commonly quantified by the economic benefits they provide, the IPBES proposes that NCP can be valued also by their intrinsic biophysical and socio-cultural benefits, such as species number or water yield. This broadens the possibilities for quantifying the benefits of landscapes, for comparing scenarios, and finally for designing and choosing desirable development pathways for a given landscape context.

In this chapter, we present the concepts and discuss how land sparing and land sharing strategies can successfully achieve FLR goals in the context of ecosystem services and the newly proposed NCP. First, we discuss pros and cons of both approaches, then we present case studies for land sparing and land sharing in the context of FLR, and, finally, we discuss implications of both approaches for decision-making.

Pros and cons of land sharing–sparing approaches

Although the debate on land sparing and land sharing has gained increased attention with the landmark publication of Green *et al.* (2005), the discussion on the concepts behind these terms has been going on much longer. For example, land sparing was argued within the considerations of the benefits of the Green Revolution, wherein the yield-improving technologies applied in the 1960s could free more land for conservation (Borlaug, 2007).

The land sparing approach argues that high-yielding agriculture can provide food for the increasing global population and diminish the demand for land clearing, allowing areas to be spared for nature conservation (Balmford *et al.*, 2012) (Figure 7.1). Under this approach, pastures and cropping fields should occupy land with high production potential, freeing

Figure 7.1 An example of a land sparing approach at São Luiz do Paraitinga municipality, state of São Paulo, Brazil. Pasture is the predominant land use, and a large area is dedicated exclusively to conservation.

Source: photo © Márcio Rangel – IIS.

other areas for nature conservation. Maximizing agricultural productivity within land sparing can be achieved, for instance, through sustainable intensification on already converted agricultural lands; a solution promoted by various researchers and policymakers (Smith *et al.*, 2010; Godfray *et al.*, 2010; Licker *et al.*, 2010; Foresight, 2011). The concept of sustainable intensification means, in essence, 'producing more food from the same area of land while reducing its environmental impacts' (Royal Society of London, 2009) by means of resource conservation and improvements in land management (Tilman *et al.*, 2002; Herrero *et al.*, 2010).

The land sharing approach, on the other hand, argues for the integration of food production and nature conservation through low-input systems and wildlife-friendly agriculture (e.g. organic farming and agroforestry systems) embedded in a mosaic landscape composed of different forest types, agricultural fields and pasturelands (Perfecto and Vandermeer, 2010; Phalan *et al.*, 2011; Balmford *et al.*, 2015; Kremen, 2015; Renwick and Schellhorn, 2016; Mertz and Mertens, 2017) (Figure 7.2). It is closely linked with traditional and more extensive agricultural systems that rely on

Figure 7.2 An example of a land sharing approach at Sítio do Arruda in São Miguel do Iguaçu, Paraná, Brazil. Agroforestry systems where shaded coffee is cultivated under the canopy of native tree species and native and exotic fruit trees can provide shelter for local biodiversity at the same time as producing food.

Source: photo © Fernanda Tubenchlak – IIS.

in situ nutrient cycling rather than external inputs (i.e. chemical fertilizers) to maintain productivity, and usually promotes a more permeable landscape matrix. This approach accommodates technological and financial constraints, often a barrier to intensification.

Maximizing biodiversity conservation is a common goal to both sides of the debate, yet the groups of species conserved in each approach may differ depending on the species' ecological requirements and the landscape features promoted by each approach. It is argued that land sparing tends to favour specialist species that require large habitat areas to persist in the long term, while generalist species that require small habitat areas may thrive under land sharing systems. The land sparing approach aims to maintain forest patches within the landscape as biodiversity reservoirs, focusing on conserving patches with high ecological integrity. Based on the premise that continuous forest habitats have higher ecological value than the same area split into smaller habitat patches of forest (Fahrig, 2017), land sparing approaches promote the protection of large patches and, therefore, the conservation of forest-specialist species. Conversely, the land sharing approach focuses on the quality of the matrix rather than on the patch quality and size. Based on the premise that a permeable matrix increases the connectivity between patches, minimizing the negative effects of habitat fragmentation (Crouzeilles *et al.*, 2013) and favouring metapopulation structures (Perfecto and Vandermeer, 2010), this approach promotes a mosaic landscape that may benefit generalist over forest-specialist species.

The groups of species that are most efficiently conserved by each approach are still under debate. Recently, Fahrig (2017) conducted an extensive review of habitat fragmentation effects on species reactions and found more positive responses in landscapes where the habitat is fragmented into several patches than in those where the habitat is concentrated in a single large patch. Species richness, for example, was higher in several small patches than in a single large patch, independently of its taxonomic group, species requirements (generalist or specialist), threat status (International Union for Conservation of Nature (IUCN)) or region (tropical or temperate). Thus, Fahrig (2017) suggests that land sharing may provide higher ecological value than land sparing. Most studies on the topic, however, point towards land sparing as a better solution for biodiversity conservation for certain species (e.g. Phalan *et al.*, 2011; Chandler *et al.*, 2013; Gilroy *et al.*, 2014), and land sharing as a better option for providing ecosystem services other than biodiversity (e.g. Chappell *et al.*, 2009; Mastrangelo and Gavin, 2012; Pywell *et al.*, 2012; Kremen, 2015). However, there is no consensus in the literature on which approach is better in the long term.

It is also critical to note that both approaches may also have negative consequences at the landscape level. The land sparing approach requires intensifying crop production to enhance productivity, often using chemical fertilizers and pesticides, which can negatively impact agrobiodiversity and pollination services and increase pollution (Tscharntke *et al.*, 2012; van Lexmond *et al.*, 2015; Law *et al.*, 2015; IPBES, 2016). Such conventional agricultural intensification (as opposed to 'sustainable intensification' described earlier), when done on a large scale, threatens smallholder farming systems that rely on biodiversity (planned and associated) and ecological processes to support their yields (e.g. Chappell *et al.*, 2009; Mastrangelo and Gavin, 2012; Pywell *et al.*, 2012; Kremen, 2015). It also implies important cultural and political challenges, given that it demands land owners' willingness to change their practices and large-scale planning from governments (e.g. land use zoning, technical assistance). Additionally, the spatial heterogeneity within the landscape may not always allow the conservation of large, continuous patches of forest, as patches of productive soils or favourable topography for mechanization, for example, may be scattered in the landscape (Perfecto and Vandermeer, 2010).

Despite the common goal of land sparing and sharing to reduce deforestation, both approaches imply risks of rebound effects that could lead to further expansion of agricultural lands at the expense of forests (Rudel *et al.*, 2009; Goulart *et al.*, 2016; Phalan *et al.*, 2016). Land sharing may, purposely or not, favour low-yield agricultural systems, which may require the cultivation of larger areas to meet the growing demand for food. In turn, land sparing relies on the use of ever-improving technology for increasing yields and, consequently, profits, which may stimulate agricultural expansion to natural habitats (Law *et al.*, 2015; Mertz and Mertens, 2017; Perfecto and Vandermeer, 2010). Furthermore, increasing yields to comply with growing demand for food can be questioned, as current food production already exceeds demand and about a third of it is wasted (Tscharntke *et al.*, 2012). Thus, side-measures for reducing food waste and demand should be considered in this debate. Given that pasturelands are one of the greatest drivers of deforestation, dietary shifts towards less meat and more plant-based diets, for example, can free up land for conservation and restoration (Stehfest *et al.*, 2009; Springmann *et al.*, 2016; Clark and Tilman, 2017). Differences between land sparing and sharing are summarized in Table 7.1.

Case studies for land sparing and land sharing in the context of forest landscape restoration

Sustainable intensification for land sparing

At the centre of the land sparing approach is the evidence showing that current agricultural productivity is below potential yields (FAO/IIASA, 2010; Licker *et al.*, 2010; Foresight, 2011; Mueller *et al.*, 2012; Strassburg *et al.*, 2014). This has been demonstrated at both global and local scales.

Table 7.1 Main differences between land sharing and land sparing approaches. The text across both columns indicates similar gains and losses between the strategies

Land sparing	Land sharing
Proposes the segregation of food production and nature conservation (at farm and landscape scales)	Proposes the integration of food production and nature conservation (at farm and landscape scales)
Defends high-yielding agriculture on a part of land which would allow sparing natural areas for conservation	Defends wildlife-friendly agriculture, despite potential lower yields
Characterized by homogeneous landscapes, where part of land is reserved for production and the non-productive part of land is reserved for nature conservation	Characterized by more heterogeneous mosaic landscapes composed of different forest types, agricultural fields and pasturelands
Argues that large patches concentrated in non-productive parts of the landscape can guarantee biodiversity conservation	Argues that several patches dispersed in the landscape can improve the matrix quality and guarantee biodiversity conservation
Associated with the implementation of monocultural cropping systems	Associated with the implementation of organic farming and agroforestry systems
Favours the conservation of specialist species with a low tolerance for disturbances	Favours the conservation of generalist species with a high tolerance for disturbances
Favours the conservation of large patches of natural vegetation	Favours the conservation of small patches of natural vegetation within agricultural landscapes
Favours the provision of a narrow range of ecosystem services	Favours the provision of a broader range of ecosystem services
Associated with hi-tech agricultural practices and large-scale agriculture	Associated with traditional production systems and smallholder farming
Involves the risk of rebound effect, whereby the use of ever-improving technology for increasing yields and profits may stimulate agricultural expansion into forests	Involves the risk of rebound effect, whereby the low-yields can require that larger areas are cultivated to meet growing demand for food and ensure profitability at the farm level
Few implementation examples of land sparing/sharing have been reported in the literature, and their results represent specific geographic regions and species	
The strategies do not take into account site history, social value, surrounding landscape and scale influence	
Drivers and expected outcomes of land sparing/sharing policies are not fully understood; clear guidelines for policymakers are lacking	
Neither strategy guarantees that the excess land will be used for habitat conservation or restoration	

For instance, Mueller *et al.* (2012) showed that closing the yield gaps by increasing productivity between 45% and 70% for most crops can be attained through changes in management practices. Besides increasing productivity, such practices may also improve environmental conditions and prevent degradation by applying zero or reduced tillage, mulches and cover crops (Godfray *et al.*, 2010) and using rotational grazing in pasturelands (Latawiec *et al.*, 2015).

In Brazil, a strategy combining sustainable intensification with land sparing has been proposed for extensive pasturelands characterized by low productivity (Strassburg *et al.*, 2014). Extensive cattle ranching systems in Brazil have historically required new patches of forest to be cleared to sustain cattle herds, as soil became rapidly degraded. By increasing the productivity of pasturelands, in most cases by applying rotational grazing or a mix of forage grasses with legumes (while taking into account sustainable carrying capacity), it is possible to reconcile future demands for agricultural products and spare land for restoration (Strassburg *et al.*, 2014).

Sustainable intensification has also been proposed as a key strategy for land sparing when planning restoration in the Brazilian Atlantic Forest biome (Latawiec *et al.*, 2016). The Atlantic Forest is a highly threatened, deforested and fragmented biome in Brazil, referred to as the 'hottest of the hotspots' on account of its high biodiversity value (Laurance, 2009). Espírito Santo, one of the states in the Atlantic Forest biome, foresees a 284,000 ha expansion of areas devoted to agricultural crops and a 400,000 ha expansion of forest plantations in the next decades (PEDEAG, 2008). In addition, there is a state government goal to restore 236,000 ha of forest by 2025, supported by the State Institute for Environment and Hydrological Resources (IEMA – Instituto Estadual do Meio Ambiente e Recursos Hídricos) and the Agroforestry and Forestry Defense Institute (IDAF – Instituto de Defesa Agropecuária e Floresta). By increasing the cattle ranching productivity in the Espírito Santo state from 1.42 million animal units (AU) (0.74 AU ha^{-1}, corresponding to only 27% of the estimated capacity) to 5.29 million AU (2.77 AU ha^{-1}), enough land may be spared for restoration to meet government plans (Latawiec *et al.*, 2014a). If implemented correctly and inclusively with respect to all stakeholders, this strategy could be extrapolated across Brazil, as recently demonstrated for another biodiversity hotspot in the country – the Cerrado biome (Strassburg *et al.*, 2017). This is especially timely given national plans for reforestation. Sustainable intensification of pasturelands can also result in positive returns to land owners, small farmers, traders and ultimately governments, through increased tax returns and multiplier effects on the economy.

Serious barriers exist to widespread sustainable intensification coupled with land sparing. Major bottlenecks to sustainable increases in productivity include labour scarcity, bureaucratic credit access and lack of adequate technical assistance (in numbers as well as in quality, since agrochemical vendors assume the role of 'technical assistants' to a significant number of farmers; Latawiec *et al.*, 2017). In addition, there are disagreements on how to assess 'agricultural sustainability' in practice (Latawiec *et al.*, 2017). Moreover, complementary policies such as territorial planning, monitoring, tenure security and improved law enforcement must be put in place to avoid undesirable outcomes such as leakage (unintended displacement of production activities) and rebound effects. These aspects are especially important in developing countries where lack of secure or legal land tenure and limited law enforcement pose a challenge to sustainable intensification.

Land sharing at regional scale – the case of Paranapanema (São Paulo State, Brazil)

An interesting example of land sharing can be drawn from the Paranapanema case study in the state of São Paulo. In the mid-1990s, 'The Rural Landless Workers Movement' (MST – Movimento dos Trabalhadores Rurais Sem Terra, in Portuguese) resettled more than 3000 families in the Pontal do Paranapanema, in the area of the 'Reserva do Pontal' National Park designated to protect the highly threatened Atlantic Forest ecosystem and the endangered black lion tamarin (*Leontopithecus chrysopygus*), endemic to the region (Hart *et al.*, 2016). This area was characterized by deforestation for cattle ranching and logging for timber production, involving serious conflicts over land ownership. New settlements were allocated in the buffer zones of the park and around the remaining forest fragments to reduce conflicts with large-scale farmers engaged mainly in intensive agriculture. This allocation, however, generated concerns in the conservation community as to whether it could threaten nature conservation in the buffer zone of the park. In response to that concern, a concerted effort led by the Institute of Ecological Research (IPÊ – Instituto de Pesquisas Ecológicas) and involving a wide range of stakeholders, including the members of MST, local NGOs and public agencies, aimed to turn the challenge of the reformed land into an opportunity to support rural livelihoods by promoting sustainable agroforestry systems and creating ecological corridors to link forest fragments within an agro-environmental matrix (Wittman, 2010). This matrix included agricultural fields for fruit, timber and fuel wood mixed with maize, beans and cassava production, creating a buffer zone for wildlife reserves and new habitats for endangered Atlantic Forest species (Cullen *et al.*, 2005). Some positive outcomes of this land sharing study were the improvement of agricultural productivity and income for local communities (Cullen *et al.*, 2005; Rodrigues *et al.*, 2007), increased awareness regarding environmental management, with attention to streams and riparian forest, and diminished use of chemical pesticides, as well as the implementation of reforestation actions that broadened the buffer zone and helped to maintain the Reserve (Wittman, 2010). Furthermore, the presence of MST settlers helped to monitor the protected area, and although some timber species continue to be illegally logged, the Reserve remains intact.

Land sparing and land sharing at a country scale – the case of Suriname

The land sparing–land sharing dichotomy can be reconciled spatially. The analysis of land use in Suriname, based on modelling and field surveys, demonstrated that it is possible to combine extensive, small-scale organic farming with sustainable intensification of rice production on converted agricultural lands (Latawiec *et al.*, 2014b). By both stimulating organic land-sharing farming and increasing the efficiency of under-productive monoculture rice systems, the country may benefit from an increased value of its national agriculture, create alternative and higher incomes, create new job opportunities and achieve food security. Because organic farming may, in certain circumstances, provide lower yields (Seufert *et al.*, 2012), coupling it with intensified systems may enable sparing land for nature. This is a promising strategy, especially for countries with high levels of biodiversity preserved (Latawiec *et al.*, 2015). The issue of scale, paramount for the land sparing and land sharing debate, is further explored in the later discussion on global policies for land sharing and sparing.

Benefits of forest landscape restoration under land sparing and land sharing approaches in São Paulo State (Brazil)

The state government of São Paulo (Brazil) is currently applying the framework of The Economics of Ecosystems and Biodiversity (TEEB) to a 1.6 million hectare landscape located within the Paraíba do Sul river watershed, aiming to support public policies and initiatives for sustainable development through the quantification and valuation of ecosystem services (IIS, 2017). TEEB is a global initiative aiming to highlight the economic benefits of biodiversity, based on a framework that helps stakeholders to recognize, demonstrate and capture the economic value of ecosystem services (Sukhdev *et al.*, 2010). To evaluate whether and how a sustainable development pathway would improve the landscape value, alternative future land use scenarios were modelled and compared in terms of ecosystem services provision and economic gain. The scenarios differed mostly with respect to the allocation of forest restoration according to a land sharing or land sparing approach.

The Paraíba do Sul river watershed is located between two of the largest Brazilian cities (São Paulo and Rio de Janeiro), and due to its strategic location, the region has historically been subject to intensive land use. Originally covered by the Atlantic rainforest, the area was largely deforested from the late eighteenth century to establish coffee plantations, which were later replaced by pastures in the early twentieth century. Due to this long-term and intensive land use history, large areas within the basin had been deforested and degraded. Today, to comply with the current Brazilian environmental legislation, it is estimated that about 79,000 hectares need to be restored at the portion of the basin located in the state of São Paulo (Soares-Filho *et al.*, 2014).

The Native Vegetation Protection Law (Law 12.651/2012) states that areas along streams, springs, mountaintops and steep hills (>45 degrees) need to be covered with native vegetation (these areas are called 'Permanent Preservation Areas' – PPAs). Beyond those areas, large properties (~100 ha) need to maintain 20% of their area covered with native Atlantic Forest (this forested portion of the property is called a 'legal reserve' – 'LR'). If the LR has already been deforested, it must be restored with native vegetation either within the property or in an equivalent area in the same biome.

In the TEEB framework applied to the Paraíba do Sul River Basin, three different future scenarios were modelled: a 'business-as-usual scenario' (BAU), in which current land use trends are projected to the future and no restoration takes place; a 'legal compliance scenario' (LC), in which all restoration required by law is carried out within each property without spatial planning, including PPAs (~26,540 ha) and LRs (~52,400 ha); and a 'sustainable' scenario (SS, in which (1) the restoration of LRs is executed with spatial planning to minimize restoration costs (costs of land plus costs of restoration practices) and maximize landscape connectivity, regardless of property boundaries, and (2) part of the current agricultural areas or pastures are replaced by sustainable productive systems (agroforestry systems, silvipastoral systems and 'voisin' pastoral systems (similar to rotational grazing)), in order to increase local food production and contribute to food security (adding up to a total of ~ 48,000 ha of sustainable production systems).

As a consequence of the different criteria for allocating restoration between the LC and SS scenarios, highly divergent spatial patterns emerged regarding the projected forest cover: while on LC the restoration was homogeneously distributed across the landscape, in the SS scenario the restoration was allocated mostly on peripheral areas of the basin, where opportunity costs are lower and connectivity is higher (given the concentration of forest remnants on the periphery of the basin). In other words, restoration areas were allocated on LC according to a 'land sharing' strategy, in which small fragments of restored forests are scattered throughout the agricultural landscapes, while on SS, restoration was allocated according to a 'land sparing' strategy, with larger and more continuous forest fragments concentrated in peripheral areas. Table 7.2 summarizes the main results regarding the LC and SS scenarios and their impact on the provision of different ecosystem services (habitat availability, water yield, sediment retention, carbon sequestration and pollination).

The land sparing scenario (SS) was more beneficial than the land sharing scenario (LC) with regard to the conservation and restoration of biodiversity (indicated by the higher habitat availability for species with different

Table 7.2 Quantification of some ecosystem services provided by a landscape in southeastern Brazil (Paraíba do Sul River Watershed), where forest restoration is allocated according to a 'land sharing' (LC) or a 'land sparing' (SS) strategy. The ecosystem services modelled were habitat availability, water yield, sediment retention, carbon sequestration and pollination

Ecosystem service	Unit	Scenario	
		LC	SS
		'Land sharing'	*'Land sparing'*
Habitat availability			
Species with dispersal capacity <100 m	Difference from BAU (%)	10.0	37.0
Species with dispersal capacity <500 m	Difference from BAU (%)	23.0	36.0
Species with dispersal capacity <1000 m	Difference from BAU (%)	21.0	33.0
Species with dispersal capacity <3000 m	Difference from BAU (%)	16.0	25.0
Water yield	Difference from BAU (%)	-2.6	-3.8
Sediment retention			
Potential soil loss (USLE equation)	Difference from BAU (%)	-4.8	-4.5
Sediment export	Difference from BAU (%)	-7.6	-3.2
Carbon sequestration	Difference from BAU (%)	2.3	2.5
Pollination			
Visitation potential – perennial and semi-perennial crops	Difference from BAU (%)	36.0	64.4
Productivity – perennial and semi-perennial crops	Difference from BAU (%)	7.2	14.5
Visitation potential – annual crops	Difference from BAU (%)	20.4	-18.2
Productivity – annual crops	Difference from BAU (%)	4.7	-7.0
Costs			
Restoration costs	Thousand US$	38,982	28,197
Opportunity costs	Thousand US$	89,865	10,445

Note
The estimated costs (in US$) to implement restoration in each scenario are also presented (using the exchange rate 1 US$ = 3.24 Brazilian Reais). USLE – Universal Soil Loss Equation; BAU – 'Business as Usual' scenario

dispersal capacities), and the costs to implement restoration in the SS scenario were 70% lower than in the LC scenario (Table 7.2). On the other hand, FLR through the land sparing approach was less effective than the land sharing approach in improving other important ecosystem services, such as water provision and sediment retention (Table 7.2). Carbon sequestration was slightly higher in SS than in LC due to the expansion of agroforestry systems in SS. For pollination, both scenarios demonstrated benefits and disadvantages, but given the fact that MSP involved other changes in land use (e.g. the expansion of agroforestry systems, which also changes pollinator abundance, and can be considered a land sharing strategy), these differences cannot be directly ascribed to different restoration strategies. In short, both strategies, land sharing (LC) and land sparing (SS), when applied to the context of the Paraíba do Sul River watershed, showed advantages and shortcomings related to the provision of different ecosystem services.

Global policies for land sharing and sparing

The Bonn Challenge on FLR, the New York Declaration on Forests, the 20 × 20 in Latin America and the AFR100 in Africa are international commitments that seek to restore up to 350 million hectares of forests on degraded and deforested lands by 2030 and support people's sustainable livelihoods (Chazdon *et al.*, 2017). FLR may include different interventions to enhance ecological functionality and human wellbeing in degraded landscapes, including land sparing and land sharing approaches. The approach to be predominantly used will depend on the scale at which the intervention is planned and the landscape context. Land sparing could be favoured in landscapes dominated by less intense land uses, where marginal lands are available for restoration and where it is possible to accommodate the potentially displaced activities through improvements in productivity (e.g. Latawiec *et al.*, 2015). Although land sharing could also be applied in these landscapes, this approach could be more efficient in landscapes where it is possible to reach a minimum forest cover to support biodiversity. For example, in the Atlantic Forest, 30% of forest remnants is the minimum required forest cover to sustain functioning communities of plants and animals (also called 'community integrity'; Banks-Leite *et al.*, 2014). Thus, it may be that landscapes with less than 30% forest cover and with low potential for sustainable intensification should focus on the land sharing approach to achieve restoration goals.

But how can one identify when and where each approach is most viable and effective in FLR programmes? We suggest seven best practices that need to be considered when addressing this question. First, define the spatial scale of analysis based on the desired FLR targets. Second, promote participatory planning and social mapping strategies for capturing land owners', decision-makers' and key stakeholders' views and requirements.

Third, explicitly account for the socio-economic and environmental drivers of FLR in each specific landscape context or region. Fourth, include spatial intelligence by identifying the most cost-effective areas for each approach based on the socio-economic and environmental contexts of the landscape or region. Fifth, highlight the trade-offs and synergies among different outcomes for each approach. Sixth, validate the results with land owners, decision-makers and key stakeholders, and promote capacity building. Seventh, promote an adaptive management process based on lessons learned from the successes and failures of already existing case studies.

Conclusions

Both land sharing and land sparing strategies have advantages and shortcomings for forest landscape restoration, with clear trade-offs for the provision of ecosystem services. They need to be thoughtfully considered, since the wide diversity of environmental and social contexts impedes the selection of a universal best solution. The best strategy to minimize trade-offs and improve the quality of agroecosystems may be to consider land sparing and land sharing as complementary, and support combinations of both in forest landscape restoration, according to the context.

A major constraint on reaching a consensus in the land sparing and land sharing debate is the lack of strong evidence supporting one strategy or the other in different socio-economic contexts. Only a few examples of implementation of these approaches have been reported in the literature, and their results are limited to specific geographic regions and/or groups of species (Bennett, 2017). The scarcity of quantitative data on the benefits and drawbacks of both approaches, along with disagreements regarding how best to quantify them, further magnifies the uncertainties (Bennett, 2017). There is also a scarcity of data on long-term effects on species persistence in landscapes that have adopted land sparing or sharing (or both) strategies (Kremen, 2015). Furthermore, some studies have not considered relevant information, such as site history, social values, surrounding landscape and scale (Renwick and Schellhorn, 2016; Bennett, 2017). Finally, the drivers and expected outcomes of land sparing and sharing policies are not fully understood, which, as a consequence, results in a lack of clear guidelines for policymakers (Mertz and Mertens, 2017).

The decision on whether land sparing or land sharing is better for nature and human wellbeing will depend on a multitude of direct, indirect, and often cumulative and uncertain consequences in the landscape. Therefore, a 'one size fits all' solution may not be realistic, and every restoration approach should allow adaptive management to ensure the success of landscape management in a wide range of environmental and social contexts. Quantifying the biophysical and economic benefits of a range of ecosystem services and NCP from rural landscapes can help in the design of restoration strategies that suit different landscape contexts. Therefore, FLR that embraces the concept of a broader and inclusive approach of NCP at the heart of the solution is a promising way forward.

Acknowledgements

We thank Ana Castro from the International Institute for Sustainability for her help with reference formatting.

References

Balmford, A., Green, R. and Phalan, B. (2012). 'What conservationists need to know about farming'. *Proceedings of the Royal Society B*, available at doi:10.1098/rspb.2012.0515 (accessed 19 June 2018).

Balmford, A., Green, R. and Phalan, B. (2015). 'Land for food & land for nature?' *Daedalus*, vol 144, pp. 57–75. doi:10.1162/DAED_a_00354.

Banks-Leite, C., Pardini, R., Tambosi, L. R., Pearse, W. D., Bueno, A. A., Bruscagin, R. T., Condez, T. H., Dixo, M., Igari, A. T., Martensen, A. C. and Metzger, J. P. (2014). 'Using ecological thresholds to evaluate the costs and benefits of set-asides in a biodiversity hotspot'. *Science*, vol 345, no 6200, pp. 1041–1045.

Bennett, E. M. (2017). 'Changing the agriculture and environment conversation'. *Nature Ecology & Evolution*, vol 1, pp. 1–2, doi:10.1038/s41559-016-0018.

Borlaug, N. (2007). 'Feeding a hungry world'. *Science*, vol 318, pp. 115–1062.

Chandler, R. B., King, D. I., Raudales, R., Trubey, R., Chandler, C. and Arce Chávez, V. J. (2013). 'A smallscale land-sparing approach to conserving biological diversity in tropical agricultural landscapes'. *Conservation Biology*, vol 27, pp. 785–795.

Chappell, M. J., Vandermeer, J., Badgley, C. and Perfecto, I. (2009). 'Wildlife-friendly farming vs land sparing'. *Frontiers in Ecology and the Environment*, vol 7, pp. 183–184.

Chazdon, R. L., Brancalion, P. H. S., Lamb, D., Laestadius, L., Calmon, M. and Kumar, C. (2017). 'A policy-driven knowledge agenda for global forest and landscape restoration'. *Conservation Letters*, vol 10, no 1, pp. 125–113.

Clark, M. and Tilman, D. (2017). 'Comparative analysis of environmental impacts of agricultural production systems, agricultural input efficiency, and food choice'. *Environmental Research Letters*, vol 12, p. 064016.

Crouzeilles, R., Lorini, M. L. and Grelle, C. E. V. (2013). 'The importance of using sustainable use protected areas for functional connectivity'. *Biological Conservation*, vol 159, pp. 450–457.

Cullen, L., Alger, K. and Rambaldi, D. M. (2005). 'Land reform and biodiversity conservation in Brazil in the 1990s: Conflict and the articulation of mutual interests'. *Conservation Biology*, vol 19, pp. 747–755.

Fahrig, L. (2017). 'Ecological response to habitat fragmentation per se'. *Annual Review of Ecology, Evolution, and Systematics*, vol 48, pp. 1–23.

FAO (Food and Agriculture Organization of the United Nations)/IIASA (International Institute for Applied Systems Analysis). (2010). *FAO/IIASA Global Agro-ecological Assessment Study*. GAEZ online database.

Foresight. (2011). *The Future of Food and Farming 2011.* Final Project Report. The Government Office for Science, London.

Gilroy, J. J., Edwards, F. A., Medina Uribe, C. A., Haugaasen, T. and Edwards, D. P. (2014). 'Surrounding habitats mediate the trade-off between land-sharing and land-sparing agriculture in the tropics'. *Journal of Applied Ecology*, vol 51, pp. 1337–1346.

Godfray, H. C. J., Beddington, J. R., Crute, I. R., Haddad, L., Lawrence, D., Muir, J. F., Pretty, J., Robinson, S., Thomas, S. M. and Toulmin, C. (2010). Food security: The challenge of feeding 9 billion people. *Science*, vol 327, pp. 812–818.

Goulart, F. F., Perfecto, I., Vandermeer, J., Boucher, D., Chappell, M. J., Fernandes, G. W., Scariot, A., Silva, M. C., Oliveira, W., Neville, R., Moore, J., Bustamante, M., Carvalho, S. R. and Soares-Filho, B. (2016). 'Emissions from cattle farming in Brazil'. *Nature Climate Change*, vol 6, pp. 893–894. doi:10.1038/nclimate3123.

Green, R. E., Cornell, S. J., Scharlemann, J. P. W. and Balmford, A. (2005). 'Farming and the fate of wild nature'. *Science*, vol 307, no 5709, pp. 550–555. doi:10.1126/science.1106049.

Hart, A. K., McMichael, P., Milder, J. C. and Scherr, S. J. (2016). 'Multi-functional landscapes from the grassroots? The role of rural producer movements'. *Agriculture and Human Values*, vol 33, pp. 305–322.

Herrero, M., Thornton, P. K., Notenbaert, A. M., Wood, S., Msangi, S., Freeman, H. A., Bossio, D., Dixon, J., Peters, M., Van De Steeg, J., Lynam, J., Rao, P. P., Macmillan, S., Gerard, B., McDermott, J., Sere, C. and Rosegrant, M. (2010). 'Smart investments in sustainable food production: Revisiting mixed crop–livestock systems'. *Science*, vol 327, pp. 822–825.

IIS (International Institute of Sustainability). (2017). Valoração socioeconômica e ecológica dos serviços ecossistêmicos da Bacia do Rio Paraíba do Sul porção paulista (TEEB-SP).

IPBES. (2016). 'The Assessment Report on Pollinators, Pollination and Food Production of the Intergovernmental Science-Policy Platform on Biodiversity and Ecosystem Services'. doi:ISBN: 978-92-807-3568-0.

Janishevski, L., Santamaria, C., Gidda, S. B., Cooper, H. D. and Brancalion, P. H. S. (2015). 'Ecosystem restoration, protect areas and biodiversity conservation'. *Unasylva*, vol 66, pp. 19–28.

Kremen, C. (2015). 'Reframing the land-sparing/land-sharing debate for biodiversity conservation'. *Annals of the New York Academy of Sciences*, vol 1355, no 1, pp. 52–76. doi:10.1111/nyas.12845.

Latawiec, A. E., Crouzeilles, R., Brancalion, P. H., Rodrigues, R. R., Sansevero, J. B., Santos, J. S. D., Mills, M., Nave, A. G. and Strassburg, B. B. (2016). 'Natural regeneration and biodiversity: A global meta-analysis and implications for spatial planning'. *Biotropica*, vol 48, no 6, pp. 844–855.

Latawiec, A. E., Strassburg, B. B. N., Valentim, J. F., Ramos, F. and Alves-Pinto, H. N. (2014a). 'Intensification of cattle ranching production systems: Socioeconomic and environmental synergies and risks in Brazil'. *Animal*, vol 8, pp. 1255–1263.

Latawiec, A. E., Strassburg, B. B. N., Rodriguez, A. M., Matt, E., Nijbroek, R. and Silos, M. (2014b). 'Suriname: Reconciling agricultural development and conservation of unique natural wealth'. *Land Use Policy*, vol 38, pp. 627–636.

Latawiec, A. E., Strassburg, B. B. N., Brancalion, P. H. S., Rodrigues, R. R. and Gardner, T. (2015). 'Creating space for large-scale restoration in tropical agricultural landscapes'. *Frontiers in Ecology and the Environment*, vol 13, pp. 211–218.

Latawiec, A. E., Strassburg, B. B. N., Silva, D., Alves-Pinto, H. N., Feltran-Barbieri, R., Castro, A., Iribarrem, A., Rangel, M. C., Kalif, K. A., Gardner, T. and Beduschi, F. (2017). 'Improving land management in Brazil: A perspective from producers'. *Agriculture, Ecosystems & Environment*, vol 240, pp. 276–286.

Laurance, W. F. (2009). 'Conserving the hottest of the hotspots'. *Biological Conservation*, vol 142, p. 113.

Law, E. A., Meijaard, E., Bryan, B. A., Mallawaarachchi, T., Koh, L. P. and Wilson, K. A. (2015). 'Better land-use allocation outperforms land sparing and land sharing approaches to conservation in Central Kalimantan, Indonesia'. *Biological Conservation*, vol 186, pp. 276–286. doi:10.1016/j.biocon.2015.03.004.

Licker, R., Johnton, M., Foley, J. A., Barford, C., Kucharik, C. J., Monfreda, C. and Ramankutty, N. (2010). 'Mind the gap: How do climate and agricultural management explain the "yield gap" of croplands around the world?' *Global Ecology and Biogeography*, vol 19, pp. 769–782.

Maginnis, S. and Jackson, W. (2007). 'What is FLR and how does it differ from current approaches?' in: J. Reitbergen-McCracken, S. Maginnis and A. Sarre (eds) *The Forest Landscape Restoration Handbook*. Earthscan, London, UK. pp. 5–20.

Mansourian, S., Vallauri, D. and Dudley, N. (2005). *Forest Restoration in Landscapes: Beyond Planting Trees*. Springer, New York.

Mastrangelo, M. E. and Gavin, M. C. (2012). 'Trade-offs between cattle production and bird conservation in an agricultural frontier of the Gran Chaco of Argentina'. *Conservation Biology*, vol 26, pp. 1040–1051.

Mertz, O. and Mertens, C. F. (2017). 'Land sparing and land sharing policies in developing countries – drivers and linkages to scientific debates'. *World Development*, vol 98, pp. 523–535.

Mueller, N. D., Gerber, J. S., Johnston, M., Ray, D. K., Ramankutty, N. and Foley, J. A. (2012). 'Closing yield gaps through nutrient and water management'. *Nature*, vol 490, pp. 254–257.

Pascual, U., Balvanera, P., Díaz, S., Pataki, G., Roth, E., Stenseke, M., Watson, R. T., Dessane, E. B., Islar, M., Kelemen, E., Maris, V., Quaas, M., Subramanian, S. M., Wittmer, H., Adlan, A., Ahn, S. E., Al-Hafedh, Y. S., Amankwah, E., Asah, S. T., Berry, P., Bilgin, A., Breslow, S. J., Bullock, C., Cáceres, D., Daly-Hassen, H., Figueroa, E., Golden, C. D., Gómez-Baggethun, E., González-Jiménez, D., Houdet, J., Keune, H., Kumar, R., Ma, K., May, P. H., Mead, A., O'Farrell, P., Pandit, R., Pengue, W., Pichis-Madruga, R., Popa, F., Preston, S., Pacheco-Balanza, D., Saarikoski, H., Strassburg, B. B., van den Belt, M., Verma, M., Wickson, F. and Yagi, N. (2017). 'Valuing nature's contributions to people: The IPBES approach'. *Current Opinion in Environmental Sustainability*, vol 26, pp. 7–16.

PEDEAG – Plano Estratégico de Desenvolvimento da Agricultura: novo PEDEAG 2007–2025. (2008). *Secretaria de Estado da Agricultura, Abastecimento, Aquicultura e Pesca*. Estado do Espírito Santo, Vitória, SEAG, 284p.

Perfecto, I. and Vandermeer, J. (2010). 'The agroecological matrix as alternative to the land-sparing/agriculture intensification model'. *Proceedings of the National*

Academy of Sciences of the United States of America, vol 107, no 13, pp. 5786–5791. doi:10.1073/pnas.0905455107.

Phalan, B., Onial, M., Balmford, A. and Green, R. E. (2011). 'Reconciling food production and biodiversity conservation: Land sharing and land sparing compared'. *Science*, vol 333, no 6047, pp. 1289–1291. doi:10.1126/science.1208742.

Phalan, B., Green, R. E., Dicks, L. V., Dotta, G., Feniuk, C., Lamb, A., Strassburg, B. B. N., Williams, D. R., Ermgassen, E. K. H. J. Z. and Balmford, A. (2016). 'How can higher-yield farming help to spare nature?' *Science*, vol 351, no 6272, pp. 450–451.

Pywell, R. F., Heard, M. S., Bradbury, R. B., Hinsley, S., Nowakowski, M., Walker, K. J. and Bullock, J. M. (2012). 'Wildlife-friendly farming benefits rare birds, bees and plants', *Biology Letters*, vol 8, pp. 772–775.

Renwick, A. and Schellhorn, N. (2016). 'A perspective on land sparing versus land sharing'. In: *Learning from Agri-Environment Schemes in Australia: Investing in Biodiversity and Other Ecosystem Services on Farms*. ANU Press, Canberra, A.C.T., Australia, pp. 117–126.

Rodrigues, E. R., Cullen Jr, L., Beltrame, T. P., Moscogliato, A. V. and da Silva, I. C. (2007). 'Avaliaçao econômica de sistemas agroflorestais implantados para recuperação de reserva legal no Pontal do Paranapanema, São Paulo'. *Revista Árvore*, vol 31, pp. 941–948.

Royal Society of London. (2009). *Reaping the Benefits: Science and the Sustainable Intensification of Global Agriculture*.

Rudel, T. K., Schneider, L. Uriarte, M. *et al.* (2009). 'Agricultural intensification and changes in cultivated areas, 1970–2005'. *Proceedings of the National Academy of Sciences of the United States of America*, vol 106, pp. 20675–20680.

Sayer, J. A. (2009). 'Reconciling conservation and development: Are landscapes the answer?' *Biotropica*, vol 41, pp. 649–652.

Sayer, J., Sunderland, T., Ghazoul, J., Pfund, J. L., Sheil, D., Meijaaed, E., Venter, M., Boedhihartono, A. K., Daay, M., Garcia, C., Oosten, C. V. and Buck, L. E. (2013). 'Ten principles for a landscape approach to reconciling agriculture, conservation, and other competing land uses'. *Proceedings of the National Academy of Sciences of the United States of America*, vol 110, pp. 8349–8356.

Scherr, S. J. and McNeely, J. A. (2008). 'Biodiversity conservation and agricultural sustainability: Towards a new paradigm of "ecoagriculture" landscapes'. *Philosophical Transactions of the Royal Society B: Biological Sciences*, vol 363, no 1491, pp. 477–494.

Seufert, V., Ramankutty, N. and Foley, J. A. (2012). 'Comparing the yields of organic and conventional agriculture'. *Nature*, vol 485, no 7397, p. 229.

Smith, P., Gregory, P. J., Van Vuuren, D., Obersteiner, M., Havlik, P., Rounsevell, M., Woods, J., Stehfest, E. and Bellarby, J. (2010). 'Competition for land'. *Philosophical Transactions of the Royal Society B: Biological Sciences*, vol 365, pp. 2941–2957.

Soares-Filho, B., Rajão, R., Macedo, M., Carneiro, A., Costa, W., Coe, W., Coe, M., Rodrigues, H. and Alencar, A. (2014). 'Cracking Brazil's forest code'. *Science*, vol 344, pp. 363–364.

Springmann, M., Godfray, H. C., Rayner, M. and Scarborough, P. (2016). 'Analysis and valuation of the health and climate change cobenefits of dietary change'. *Proceedings of the National Academy of Sciences of the United States of America*, vol 113, no 15, pp. 4146–4151.

Stehfest, E., Bouwman, L., van Vuuren, D., den Elzen, M., Eickout, B. and Kabat, P. (2009). 'Climate benefits of changing diet'. *Climatic Change*, vol 95, no 1, pp. 83–102.

Strassburg, B. B. N., Latawiec, A. E., Barioni, L. G., Nobre, C. A., da Silva, V. P., Valentim, J. F., Vianna, M. and Assad, E. D. (2014). 'When enough is enough: Improved use of current agricultural lands could meet demands and spare nature in Brazil'. *Global Environmental Change*, vol 28, pp. 84–97.

Strassburg, B. B. N., Brooks, T., Feltran-Barbieri, R., Iribarrem, A., Crouzeilles, R., Loyola, R., Latawiec, A. E., Oliveira Filho, F. J. B., Scaramuzza, C. A. M., Scarano, F. R., Soares-Filho, B. and Balmford, A. (2017). 'Moment of truth for the Cerrado hotspot'. *Nature Ecology & Evolution*, vol 1, no 4, p. 99.

Sukhdev, P., Wittmer, H., Schröter-Schlaack, C., Nesshöver, C., Bishop, J., Brink, P. T., Gundimeda, H., Kumar, P. and Simmons, B. (2010). 'The economics of ecosystems and biodiversity: Mainstreaming the economics of nature: A synthesis of the approach, conclusions and recommendations of TEEB' (No. 333.95 E19). United Nations Environment Programme, Geneva.

Tilman, D., Cassman, K. G., Matson, P. A., Naylor, R. and Polasky, S. (2002). 'Agricultural sustainability and intensive production practices'. *Nature*, vol 418, pp. 671–677.

Tscharntke, T., Clough, Y., Wanger, T. C., Jackson, L., Motzke, I., Perfecto, I., Vandermeer, J. and Whitbread, A. (2012). 'Global food security, biodiversity conservation and the future of agricultural intensification'. *Biological Conservation*, vol 151, no 1, pp. 53–59.

van Lexmond, M. B., Bonmatin, J. M., Goulson, D. and Noome, D. A. (2015). 'Worldwide integrated assessment on systemic pesticides: Global collapse of the entomofauna: Exploring the role of systemic insecticides'. *Environmental Science and Pollution Research International*, vol 22, no 1, pp. 1–4.

Wittman, H. (2010). 'Agrarian reform and the environment: Fostering ecological citizenship in Mato Grosso, Brazil'. *Canadian Journal of Development Studies/ Revue canadienne d'études du développement*, vol 29, pp. 281–298.

8 Linking forest conservation and food security through agroecology

Insights for forest landscape restoration

Kristina Van Dexter and Ingrid Visseren-Hamakers

Introduction

The recent international emphasis on various landscape approaches, including forest landscape restoration (FLR), has stimulated discussions on the use of integrated policy approaches to the adaptation and mitigation potential of agriculture, forests and land use change (Grau and Aide, 2008; Reed *et al.*, 2016; Bastos Lima *et al.*, 2017). There has been increasing attention to integrated, multi-objective, cross-sectoral management of rural landscapes in order to reconcile growing demands for food, (bio-)energy, biodiversity conservation and addressing climate change in a context of conflicts over land, water and other natural resources (Sayer *et al.*, 2013).

Considering the interdependence of forests and farms and the growing area of mosaic landscapes where trees and forests are interspersed with permanent or semi-permanent agricultural uses (Perfecto *et al.*, 2009; Rahman *et al.*, 2016; Souza *et al.*, 2016), policy coordination across the agriculture, forestry and natural resources sectors is necessary (Reed *et al.*, 2016). Integrated decision-making processes from global to local levels help to achieve synergies and adequately address multiple objectives simultaneously, for example rural livelihoods, agriculture, water, energy, land and climate change. This is a challenging task, requiring new institutional and governance arrangements that not only empower stakeholders but also contribute to local livelihoods and the wellbeing of rural land users (Adams *et al.*, 2016). Nevertheless, in practice, current understanding of integrated landscape approaches is fragmented with little supporting evidence of their implementation (Reed *et al.*, 2015, 2016). It is fair to say that experience in successful policy integration remains scattered. Differing sectoral objectives compromise any attempt at defining a common path.

Given the growing interest in integrated landscape approaches (see e.g. Visseren-Hamakers, 2015), and the potential impacts of such initiatives, it is important to learn from practical experiences of applying integrated land use management throughout the world. FLR presents an approach to improving land use and achieving multiple objectives which could align with different policy objectives (van Oosten *et al.*, 2018).

In this chapter, we examine integrated land use approaches for FLR, with an emphasis on the tropics, where agricultural production often occurs within complex land use mosaics, and where integrated approaches have demonstrated the potential to address drivers of land use change and restore tropical landscapes for forest ecological and rural livelihoods outcomes. This is explored more fully through evaluation of a particular integrated approach – agroecology – as a pathway for restoring agro-forest landscapes in post-conflict Colombia.

Meeting diverse needs of multiple resource users at the landscape scale

Landscapes are composed of different land uses, complex land rights regimes, social organizations, institutional arrangements, and worldviews embedded within overlapping social, cultural and political histories and involving diverse sets of actors (Frost *et al.*, 2006; Rhemtulla and Mladenoff, 2007). In these multidimensional spaces, nature and culture interact, creating complex and adaptive systems that result from the coupling of social and ecological processes (Liu *et al.*, 2007; Binder *et al.*, 2013; Filotas *et al.*, 2014; Bennett *et al.*, 2015; Chazdon *et al.*, 2016) across multiple spatial and temporal scales (Young *et al.*, 2006; Schlüter *et al.*, 2012; Scholes *et al.*, 2013). Landscapes are multifunctional, dynamic and continuously evolving under the influence of political, market, social and environmental factors (Reed *et al.*, 2015, 2016). Within these spaces that do not correspond to any given political or jurisdictional boundary, there are both opportunities and challenges for policy integration.

In this context, the rights to access, use and manage land resources are subject to social norms and negotiation, framed by formal rules established by institutions. These shape land users' livelihoods as well as their ecological, social and economic relationships (Frost *et al.*, 2006). Because landscapes must fulfil a number of functions to satisfy a broader range of stakeholders holding divergent interests, this can lead to conflict and unsustainable land use. Considering restoration of forests within a landscape requires 'reorganization of agriculture, environment, and forestry sectors within national and regional governments to align land use policies in ways that promote biodiversity conservation and ecosystem service provision while addressing the needs of stakeholders who live on these landscapes' (Sayer *et al.*, 2013). Successful FLR efforts will, therefore, require collaborative decision-making, adequate representation of interests, an understanding of the landscape and of the relations between its various components, and institutions that incorporate cross-sectoral interests (Minang *et al.*, 2015).

Land tenure and rights represent an additional challenge at the scale of landscapes (Lazos-Chavero *et al.*, 2016), and particularly when considering modifications of the landscape through restoration (Mansourian, 2016). Applying and enforcing policies under insecure or unclear tenure presents further complexity. Lack of secure tenure may lead to the loss of access of local people to forest goods and agro-forest systems on which they depend, contributing to food insecurity and undermining livelihoods (Vira *et al.*, 2015).

Understanding and addressing land rights issues requires an understanding of history and of power relations (see Chapter 4 and Chapter 10, this volume). Landscapes are influenced by socio-cultural systems and different interests of community members, who rely on distinct tree species or varieties and use their gender-specific skills to manage and utilize these. Women often lack a voice in decisions about forest management and do not share equally in the benefits of forest tenure (Lawry *et al.*, 2016). For example, a study of decision-making in adoption of agroforestry in Africa (Villamor *et al.*, 2014) highlighted gaps in gendered knowledge, preferences, risk taking and access to innovation in land use. Men and women have different roles and knowledge of forests, and will derive different benefits from forest restoration, including monetary gains and cultural and subsistence uses.

Managing for multiple objectives

Food production systems are often embedded within landscapes that include mosaics of forests, forest fragments, agroforestry systems and agricultural systems (Foli *et al.*, 2014). Forested landscapes also provide fuel, fodder and products for income generation (Agrawal *et al.*, 2013; Vira *et al.*, 2015). Managing multifunctional landscapes in a way that combines food production, biodiversity conservation and maintenance of ecosystem services helps to achieve food security and support rural livelihoods. More than 1 billion people rely on forest products for food or income, and forests contribute as much as 20% to rural people's income (Wunder *et al.*, 2014).

Rural people in the tropics have a long tradition of managing their forests or partly forested landscapes: 40–80% of global food production comes from diverse smallholder agricultural systems in complex landscapes (FAO, 2016; IFAD, 2016). Evidence indicates that these forested landscapes are more adaptable, resilient and diverse than monocrop or plantation production systems (Vira *et al.*, 2015). Maintaining diversity in agricultural production systems leads to increased resilience to changing socio-economic and market contexts (Arnold *et al.*, 2011). Social, political, economic and environmental factors interact with each other and influence the role of forest and tree-based systems for food security and livelihoods (Chowdhury and Turner, 2006; Souza *et al.*, 2016; Tschakert *et al.*, 2007; Vira *et al.*, 2015).

The role of agroecology and agroforestry in integrated land use strategies

In order to be environmentally sustainable, integrated land use approaches must take into account the diverse conditions in which smallholders live, rely on local resources, and be based on local and indigenous knowledge (Altieri, 2002). Heterogeneity among rural land users – including access to productive land, land tenure, livelihood diversification strategies, risk taking, knowledge bases, food needs, and benefits derived from forests – affects their land use choices (Tschakert *et al.*, 2007; Bullock and King, 2011; Lazos-Chavero *et al.*, 2016; Souza *et al.*, 2016). Integrated land use approaches seek to align ecologically based agriculture, resource-based livelihoods and ecosystem conservation.

Agroecology – the science of ecology applied to the design, development and management of agriculture – offers the potential to play an integrative land use role, provided such practices can be scaled up and supported by policy frameworks at the national level (Altieri and Farrell, 1995). Agroecological approaches, including agroforestry, which build on local land use practices and knowledge (Altieri and Farrell, 1995), can conserve agricultural biodiversity while fostering ecosystem functions (e.g. soil fertility, water conservation and pest control) to support productivity and enhance biodiversity and ecosystem services. In doing so, they can also build social and ecological resilience of rural communities and support smallholder livelihoods (Altieri and Farrell, 1995; Mbow *et al.*, 2014). Agroecological practices serve to create biological corridors between protected forests, reduce soil erosion, enhance domestic fuelwood production, and more generally reduce pressure on forests and other natural resources (Altieri and Farrell, 1995; Mbow *et al.*, 2014). Trees on farms in agroforestry systems provide additional income sources for rural farmers, can maintain or improve agricultural yields relative to monocultural agricultural systems, and enhance resilience to market or climatic shocks (Tschakert *et al.*, 2007; Reed *et al.*, 2016). There is considerable evidence that forests and on-farm trees contribute significantly to food security and nutrition through a range of environmental benefits, provision of products and social co-benefits such as increased farm income, as well as ecological restoration (Grau and Aide, 2008; Foli *et al.*, 2014; Vira *et al.*, 2015).

In the following sections, we discuss land use change dynamics of a post-conflict frontier tropical forest landscape in the Colombian Amazon in terms of historical land use and drivers of deforestation and forest degradation, and current land and agrarian reforms developed as part of peace-building efforts. We then present agroecological experiments that have emerged across the landscape, emphasizing their contribution to food sovereignty and peasants' territorial expressions and rights. Finally, we discuss challenges and opportunities for FLR in post-conflict Colombia, showing how integrating agroecology to restore ecological integrity, avoid deforestation and forest degradation, and provide alternative livelihoods for displaced farmers can contribute to a peaceful future.

Agroecology as an integrated approach to landscape restoration in post-conflict Colombia

Tropical forests worldwide are increasingly impacted by agricultural expansion, the leading driver of land cover change and deforestation globally (Geist and Lambin, 2002; Lambin *et al.*, 2003; Foley, 2005; Grau and Aide, 2008; Gibbs *et al.*, 2010). They have also been the scene of armed conflicts (Dudley *et al.*, 2002; Machlis and Hanson, 2008; Hanson *et al.*, 2009), which, along with post-conflict development during peace building, interact with land use to transform forest landscapes (Álvarez, 2001; Le Billon, 2001; McNeely, 2003; Sánchez-Cuervo and Aide, 2013; Butsic *et al.*, 2015; Ordway, 2015). These effects depend on a complex interplay of historical, social, economic and political factors characteristic of tropical forest frontiers that help to determine land and forest use (Peluso and Lund, 2011; Butsic *et al.*, 2015; Baumann and Kuemmerle, 2016; Castro-Nunez *et al.*, 2017).

Forest and agricultural land use during and after armed conflict

Over the last 30 years, expansion of the agricultural frontier has led to widespread forest loss in the Colombian Amazon (Dávalos *et al.*, 2009). Agricultural colonization fronts typically occur in conflict-affected areas, which are often associated with the production of coca crops, and followed by conversion to pasture (Etter *et al.*, 2005, 2006; Armenteras *et al.*, 2006, 2013; Dávalos *et al.*, 2009, 2011; Van Ausdal, 2009; Chadid *et al.*, 2015).

Since the mid-1990s, Colombia, with the aid of the United States, has waged a war on drugs and terrorism focused on the production of coca. Indiscriminate aerial fumigation of coca plantations with glyphosate has contributed to the degradation of the forest, soils and contamination of water, impacting agricultural production of household crops, as well as forest species that are commonly used in household consumption (Messina and Delamater, 2006; Rincón-Ruiz and Kallis, 2013). This has resulted in food insecurity and displacement of rural populations forced to abandon their lands (Gonzalez, 2000; Messina and Delamater, 2006; Rincón-Ruiz *et al.*, 2013a; Rincón-Ruiz and Kallis, 2013). In addition, the expansion of extensive cattle ranching, often undertaken as a means to establish land claims and enhance the value of marginal land affected by conflict (Dávalos *et al.*, 2014; Chadid *et al.*, 2015; Castro-Nunez *et al.*, 2017), has been a leading driver of deforestation (Rincón-Ruiz *et al.*, 2013b).

At the same time, armed conflict has had the unintended consequence of reducing development pressure in the region (Dávalos, 2001; Álvarez,

2003), and has led to forced displacement and land abandonment. In some areas, this has resulted in natural regeneration and forest expansion over the last decade (Sánchez-Cuervo and Aide, 2013), creating a complex mosaic of variable-aged secondary and primary forests (Castellanos-Castro and Newton, 2015; Baptiste *et al.*, 2017; Castro-Nunez *et al.*, 2017). These remaining primary forests are repositories of a highly diverse and endemic biota and are priorities for conservation (Álvarez, 2002).

Land inequalities are both a cause and a consequence of the armed conflict in Colombia. They have played a major role in shaping the country's historical trajectory of agrarian change, as well as the violent expression of social tensions, including those between peasants and cattle ranchers and, in more recent years, agribusinesses. This conflict has greatly contributed to the impoverishment of peasants through illegal dispossession, forced displacement of peasant agriculture, and concentration of land by drug traffickers, paramilitaries, agribusiness, the military, government and the FARC (Fuerzas Armadas Revolucionarias de Colombia) (Reyes Posada, 2009).

Agrarian and land reforms are at the centre of the peace agreement between the FARC and the Colombian government. Within the framework for the implementation of the 'Final Agreement for the Termination of the Conflict and the Construction of a Stable and Lasting Peace' (Acuerdo Final para la Terminación del Conflicto y la Construccion de una paz estable y Duradera), 'Integral Rural Reform' (Reforma Rural Integral, RRI) of the peace agreement aims to address land inequalities, including through the creation of a 'land fund' for the redistribution of land to peasants and securing land rights, as well as delimiting the agricultural frontier, including through support for alternative agriculture production and peasant reserve zones (Zonas de Reserva Campesina). Land reforms will be complemented by rural development programmes that aim to provide alternatives to coca crop cultivation, including financial and technical support for farmers' agricultural production, such as access to credit, markets and infrastructure development.

Under the rubric of rural reforms promoted by the peace agreement, agricultural legislation being pushed through Congress responds to a broader model of rural development that aims to promote modernization of the agriculture sector. Recent policies seek to promote agricultural productivity and competitiveness, including through introduced technologies and seed regulations to control production. These include 'Productive Alliances for Peace', considered a key tool for the voluntary substitution of coca crops, which promote agro-productive models linking small farmers with commodity markets and supply chains. These approaches, guided by technical and commercial imperatives of state-induced notions of agricultural productivity, often disregard technologies, practices and related knowledge applied by small farmers in agroecological production systems. Under the auspices of 'consolidating peace', such approaches effectively seek to control agrobiodiversity and related knowledge, including the flow of seeds, agro-productive practices and rural economies.

In contrast, alternative proposals advocated by peasant organizations include the promotion and protection of native seeds and associated traditional knowledge in order to strengthen the productive capacities of the peasant and community economies and stimulate technological innovation. Such proposals have arisen from peasant mobilizations and related platforms (including, most notably, the 2013 national agrarian strike), and responses to policies and regulations designed to orient agriculture towards global commodity markets, including through the introduction of seed regulations and free-trade agreements.

Peace building in an agricultural frontier: the case of Putumayo

Extending from the upper Andes to the Andean Piedmont and lowland rainforest of the Amazon, Putumayo can be characterized as a frontier landscape with important tropical forest ecosystems that has been historically affected by armed conflict and agricultural colonization. Deforestation rates in Putumayo are among the highest in the country, due to uncontrolled agricultural colonization linked with land grabbing, extensive cattle ranching and the production of coca crops. Coca cultivation and cattle ranching, linked with unstable and informal economies, are the main economic activities for impoverished small farmers in this frontier.

Poor infrastructure and lack of market access for alternative agricultural products, issues historically associated with coca production in Putumayo, can be seen as also contributing to the developing problem of cattle ranching in the region. In recent decades, extensive cattle ranching has increased in Putumayo, as small farmers transitioning from coca have turned to cattle ranching in the absence of viable economic alternatives. In this remote region, the lack of infrastructure and weak economies have been a major impediment to the production and commercialization of alternative agricultural products. In addition, land grabbing involving forest conversion to pasture with cattle has been reported in forest areas previously inaccessible due to the occupation of armed groups. These emergent land use dynamics have important implications for Putumayo's tropical forests, as extensive cattle ranching practices have been linked with forest and biodiversity loss and soil degradation.

In response to the growing expansion of extensive cattle ranching in Putumayo, farmers are being trained in agroecological land use practices, including agroforestry and silvopastoral systems aimed at forest conservation and restoration. An agreement signed between farmers and government authorities in 2017 demonstrates a renewed commitment from the government to address deforestation caused by cattle ranching. It involves the participation of farmers and cattle ranching associations in the development of conservation and reforestation activities to be implemented by farmers. These measures include silvopastoral cattle ranching systems and the commercialization of non-timber forest products (NTFPs), including planting, production and processing of acai, an Amazonian palm fruit native to Putumayo. This agreement builds on the ongoing efforts of conservation non-governmental organizations (NGOs) in alliance with local and regional cattle ranching associations, which include the development of live fences and tree nurseries with farmers. Tours of pilot farms are used to promote changes in land use practices at the farm scale and to raise awareness about farm, forest ecosystem and watershed connectivity.

Despite these efforts, there are several barriers to widespread adoption of such practices. The majority of small farmers, particularly those living in remote frontier areas, have not been involved in the dialogues and capacity-building programmes. This presents an important challenge, as successful implementation of agroecological restoration efforts will depend on the extent to which they take into account the different socio-economic factors that influence farmers' land use decisions. For those farmers who desire to make changes in their farming practices, the economic realities at the forest frontier and the lack of integrated institutional support impede actions that favour conservation. Technological packages promoted by the cattle ranching sector continue to focus on economic productivity while failing to take into account the broader social and ecological contexts affecting cattle land use among small farmers and related environmental impacts. This culture is promoted from the national level to local farmers' associations. The cattle ranching sector has little incentive or motivation to internalize a more integral approach that includes ecosystem management. As a result, farmers generally do not view the incorporation of restoration practices as contributing to their production. Moreover, there is a lack of coordination between rural development efforts being implemented under the peace agreements and restoration efforts being carried out with the cattle ranching sector.

Peacebuilding efforts seeking to stabilize the agricultural frontier, substitute coca crop production, and promote rural development and land reforms are designed to address historical land conflicts and serve as a catalyst for post-conflict reconstruction. However, post-conflict development is often associated with increased levels of forest fragmentation and biodiversity loss in areas where tropical forests and conflict overlap (Álvarez, 2001; Le Billon, 2001; McNeely, 2003; Milburn, 2014). Rural development and land reforms will determine how land is used and allocated, and landscape transformation is to be expected as forest frontiers become accessible to economic interests such as agricultural development and cattle ranching, and as farmers displaced by the conflict return to their land (Dávalos, 2001; Álvarez, 2003; Baptiste *et al.*, 2017; Negret *et al.*, 2017). Programmes that support displaced farmers to return to their land, and provide subsidies for agricultural production through productive alliances that link farmers with global commodity markets, are likely to drive a rapid development of agriculture in regions that were previously inaccessible due to the presence of armed groups (Negret *et al.*, 2017). Indeed, it can be argued that conflict has helped to maintain forests in terms of their extent and quality (Álvarez, 2001; Le Billon, 2001; McNeely, 2003). Post-conflict land and agrarian reforms thus present new challenges to communities of farmers affected by prolonged conflict.

Agroecology as a pathway for restoring post-conflict landscapes

Across Putumayo, agroecological farming initiatives have emerged as alternatives to both coca cultivation and market-based agricultural production. A small yet growing group of dispersed farmers have turned to Amazon-based farming practices that seek to restore soils, agrobiodiversity and forests, as well as farmers' traditional knowledge. These practices acknowledge the relational ecologies of tropical agriculture, which include intimate interactions between sunlight, lunar cycles, fungi, rainfall, seeds, soils, animals, insects, trees and farmers, and seek to restore these relationships. Such approaches rely on situated and empirical knowledge and practices that are transmitted via social networks and community organizations.

Amazon-based agricultural systems, in their various forms, have a long history among indigenous communities in the region. Indigenous agricultural systems, known as *chagras*, are designed according to the structure, functioning and dynamics of tropical forest ecosystems and are embedded in communities' cultural systems. For indigenous communities, mythology of origin is fundamental to regulating interactions with nature, and, therefore, agriculture is associated with creation stories as well as stories related to each of the cultivated plants, forests and soils, and the animals associated with the landscape. Organized as polyculture and agroforestry systems patterns modelled on the ecological structure, composition and nutrient cycling of the Amazonian forest, agroecosystem experiments carried out by peasants on their farms include a high diversity of plant species and varieties (e.g. of yuca, banana and numerous palms). Farmers experiment with technologies learned from the surrounding ecosystem to recreate conditions for the microbial ecology of Amazonian soils and to allow the regeneration of forest, a process that involves a diversity of bird and monkey species that aid in the dispersal of seeds. To conserve Amazonian agrobiodiversity, farmers have established native seed banks and organize community workshops on seed preservation. Linking farmers to local and regional markets for their products is an important part of this transformation to ensure the sustainability and durability of their practices. Several farmers are leading initiatives to establish local farmers' markets that link agroecological farms with peasant economies.

Women's groups are supporting a growing network of women farmers in their transition from coca to the agroecological production of Amazonian cultivars in *huertas*, or home gardens. These forest-farming practices contribute to the strengthening and sharing of women's agricultural knowledge, native seeds, and the conservation of agrobiodiversity. They contribute to food security and food sovereignty starting from their households, and help to conserve and restore forest ecosystems as well as the relationship between rural communities and the land. These women meet regularly to exchange seeds and information on farming practices and technologies. Such spaces have also fostered political action and increased women's participation in the implementation of the peace agreement on particular issues related to land and agrarian reform, as well as the substitution of coca crops, helping to bridge the gender gap historically associated with issues of land access.

The strengthening and proliferation of Amazon-based agricultural land use practices, and related technologies and knowledge, depends on a 'campesino-a-campesino' (farmer-to-farmer) pedagogy (Holt-Gímenez, 2006). The emergence of agroecology schools and demonstration farms plays an important role in this process and, more broadly, supports agroecological farming in a historically decisive context for Colombian agriculture. Farmers and peasant organizations envision linking regional sites with the Instituto Agroecologico Latinoamericano (IALA), an agroecology school recently established in María Cano by the agricultural trade union FENSUAGRO (Federación Nacional Sindical Unitaria Agropecuaria). This is part of a broader effort to develop a peasant university that responds to the need for an alternative to the dominant agricultural model through efforts to develop agrobiodiverse farming systems that recognize the important role of peasants' farming practices and knowledge in ecological conservation.

Agroecological production systems can provide alternative livelihood options for farmers economically dependent on coca crops, as well as for displaced populations returning to land that had been abandoned during the conflict. Amazon-based farming practices aim not only to cultivate biodiverse agricultural systems and food sovereignty, but also to contribute towards cultivating an ecological and reparative relationality between farmers and the continuum of life in post-conflict Putumayo. Such practices are part of peasants' rights struggles and a means for constructing peace through processes that address land and agrarian conflicts, while restoring social, cultural, economic and ecological relations among marginalized communities in post-conflict landscapes. In this context, agroecology can be seen as a political expression of an emerging paradigm of food sovereignty that seeks to reorient agriculture production systems towards approaches based on land access for small farmers and ecological production practices (Altieri, 1999, 2002, 2009; Rosset *et al.*, 2006; Patel, 2009; Altieri and Toledo, 2011; Holt-Giménez and Altieri, 2012; Gliessman, 2015), which emphasize local knowledge and technologies for food production, land access for small farmers and ecological production practices, and recognize rural community, local knowledge and gender equality (Edelman *et al.*, 2014). The case of Putumayo reflects experiences across Colombia that have shown that agroecosystems contribute to social and ecological resilience (Francis *et al.*, 2003; Kremen *et al.*, 2012) and provide support for rural households in the face of economic instability (Forero, 2002; Forero *et al.*, 2013; Acevedo-Osorio *et al.*, 2015), while contributing to the conservation of peasant knowledge and biodiversity within agroecosystems and across the landscape.

Conclusion

Integrating policies to support landscape-level interventions such as FLR remains challenging in most contexts. Few successful examples exist that demonstrate how such integration can be applied.

The case study of Putumayo in post-conflict Colombia considered how agroecological and Amazonian-based farming practices can contribute to the restoration of the productive, ecological, social and political fabric of landscapes degraded by conflict and inappropriate agricultural development models that have ruptured these relationships. It also demonstrated that integrated restoration efforts are in essence political, touching upon contentious issues such as tenure and land rights, and diverging visions on development and conservation.

The case of Putumayo demonstrates, on the one hand, the conflict between centrally designed land use policies and traditional farmer practices. These policies undermine the traditions and knowledge of rural communities, whose approaches to land use have promoted forest restoration and natural regeneration. Following years of guerrilla warfare combined with the environmentally damaging 'war on drugs', there is an opportunity to support local practices that combine multiple objectives, such as food production and biodiversity conservation, through approaches such as FLR. On the other hand, a centralized approach that fails to consider local knowledge, needs and opportunities may exacerbate inequalities and promote land use practices, such as cattle ranching, that have a negative impact on land, soils and forests.

While the case of Colombia is indeed unique, it provides important lessons for efforts elsewhere, including the potential for agroecology to contribute to FLR, the value of coordination among different dispersed efforts, and the need for coherent government policies to support emerging potentially significant contributions by farmers and local communities to restoration on the ground.

Acknowledgements

The authors would like to sincerely thank Stephanie Mansourian and John Parrotta for their detailed comments on and extensive edits of the chapter.

References

Acevedo-Osorio, Á., Garavito-Morales, L., Salgado-Arroyave, D. and Gallego-Aristizábal, J. (2015). *Contribuciones de la agricultura familiar en Colombia desde el enfoque de la multifuncionalidad. Tres estudios de caso de agricultura familiar campesina e indígena (WP)*. Corporación Universitaria Minuto de Dios, Uniminuto.

Adams, C., Rodrigues, S., Calmon, M. and Kumar, C. (2016). Impacts of large-scale forest restoration on socioeconomic status and local livelihoods: What we know and do not know. *Biotropica*, 48(6), pp. 731–744.

Agrawal, A., Cashore, B., Hardin, R., Shepherd, G., Benson, C. and Miller, D. (2013). *Background Paper 1 Economic Contributions of Forests*. In United Nations Forum on Forests Tenth Session. Istanbul, Turkey.

Altieri, M. (1999). Applying agroecology to enhance productivity of peasant farming systems in Latin America. *Environment, Development and Sustainability*, 1, pp. 197–217.

Altieri, M. (2002). Agroecology: The science of natural resource management for poor farmers in marginal environments. *Agriculture, Ecosystems and Environment*, 93, pp. 1–24.

Altieri, M. (2009). Agroecology, small farms and food sovereignty. *Monthly Review*, 61(3), pp. 102–111.

Altieri, M. and Farrell, J. (1995). *Agroecology: the science of sustainable agriculture*. Boulder, CO: Westview.

Altieri, M. and Toledo, V. (2011). The agroecological revolution in Latin America: Rescuing nature, ensuring food sovereignty and empowering peasants. *Journal of Peasant Studies*, 38(3), pp. 587–612.

Álvarez, M. (2001). Could peace be worse than war for Colombia's forests? *The Environmentalist*, 21(4), pp. 305–315.

Álvarez, M. (2002). Illicit crops and bird conservation priorities in Colombia. *Conservation Biology*, 16(4), pp. 1086–1096.

Álvarez, M. (2003). Forests in the time of violence. *Journal of Sustainable Forestry*, 16(3–4), pp. 47–68.

Armenteras, D., Cabrera, E., Rodríguez, N. and Retana, J. (2013). National and regional determinants of tropical deforestation in Colombia. *Regional Environmental Change*, 13(6), pp. 1181–1193.

Armenteras, D., Rudas, G., Rodriguez, N., Sua, S. and Romero, M. (2006). Patterns and causes of deforestation in the Colombian Amazon. *Ecological Indicators*, 6(2), pp. 353–368.

Arnold, M., Powell, B., Shanley, P. and Sunderland, T. (2011). Editorial: Forests, biodiversity and food security. *International Forestry Review*, 13(3), pp. 259–264.

Baptiste, B., Pinedo-Vasquez, M., Gutierrez-Velez, V., Andrade, G., Vieira, P., Estupiñán-Suárez, L., Londoño, M., Laurance, W. and Lee, T. (2017). Greening peace in Colombia. *Nature Ecology and Evolution*, 1(4), p. 0102.

Bastos Lima, M., Visseren-Hamakers, I., Braña-Varela, J. and Gupta, A. (2017). A reality check on the landscape approach to REDD+: Lessons from Latin America. *Forest Policy and Economics*, 78, pp. 10–20.

Baumann, M. and Kuemmerle, T. (2016). The impacts of warfare and armed conflict on land systems. *Journal of Land Use Science*, 11(6), pp. 672–688.

Bennett, E., Cramer, W., Begossi, A., Cundill, G., Díaz, S., Egoh, B., Geijzen-dorffer, I.R., Krug, C.B., Lavorel, S., Lazos, E., Lebel, L., Martín-López, B., Meyfroidt, P., Mooney, H.A., Nel, J.L., Pascual, U., Payet, K., Pérez Harguin-deguy, N., Peterson, G.D., Prieur-Richard, A.-H., Reyers, B., Roebeling, P., Seppelt, R., Solan, M., Tschakert, P., Tscharntke, T., Turner, B.L. II, Verburg, P.H., Viglizzo, E.F., White, P.C.L. and Woodward, G. (2015). Linking biodiver-sity, ecosystem services, and human well-being: Three challenges for designing research for sustainability. *Current Opinion in Environmental Sustainability*, 14, pp. 76–85.

Binder, C., Hinkel, J., Bots, P. and Pahl-Wostl, C. (2013). Comparison of frame-works for analyzing social-ecological systems. *Ecology and Society*, 18(4), p. 26.

Bullock, A. and King, B. (2011). Evaluating China's Slope Land Conversion Program as sustainable management in Tianquan and Wuqi Counties. *Journal of Environmental Management*, 92(8), pp. 1916–1922.

Butsic, V., Baumann, M., Shortland, A., Walker, S. and Kuemmerle, T. (2015). Conservation and conflict in the Democratic Republic of Congo: The impacts of warfare, mining, and protected areas on deforestation. *Biological Conservation*, 191, pp. 266–273.

Castellanos-Castro, C. and Newton, A. (2015). Environmental heterogeneity influ-ences successional trajectories in Colombian seasonally dry tropical forests. *Bio-tropica*, 47(6), pp. 660–671.

Castro-Nunez, A., Mertz, O. and Sosa, C. (2017). Geographic overlaps between priority areas for forest carbon-storage efforts and those for delivering peace-building programs: Implications for policy design. *Environmental Research Letters*, 12(5), p. 054014.

Chadid, M., Dávalos, L., Molina, J. and Armenteras, D. (2015). A Bayesian spatial model highlights distinct dynamics in deforestation from coca and pastures in an Andean biodiversity hotspot. *Forests*, 6(12), pp. 3828–3846.

Chazdon, R., Brancalion, P., Laestadius, L., Bennett-Curry, A., Buckingham, K., Kumar, C., Moll-Rocek, J., Vieira, I.C.G. and Wilson, S.J. (2016). When is a forest a forest? Forest concepts and definitions in the era of forest and landscape restor-ation. *Ambio*, 45(5), pp. 538–550.

Chowdhury, R. and Turner, B. (2006). Reconciling agency and structure in empiri-cal analysis: Smallholder land use in the southern Yucatán, Mexico. *Annals of the Association of American Geographers*, 96(2), pp. 302–322.

Dávalos, L. (2001). The San Lucas mountain range in Colombia: How much con-servation is owed to the violence? *Biodiversity Conservation*, 10, pp. 69–78.

Dávalos, L., Bejarano, A. and Correa, H. (2009). Disabusing cocaine: Pervasive myths and enduring realities of a globalised commodity. *International Journal of Drug Policy*, 20(5), pp. 381–386.

Dávalos, L., Bejarano, A., Hall, M., Correa, H., Corthals, A. and Espejo, O. (2011). Forests and drugs: Coca-driven deforestation in tropical biodiversity hotspots. *Environmental Science & Technology*, 45(4), pp. 1219–1227.

Dávalos, L., Holmes, J., Rodríguez, N. and Armenteras, D. (2014). Demand for beef is unrelated to pasture expansion in northwestern Amazonia. *Biological Conservation*, 170, pp. 64–73.

Dudley, J., Ginsberg, J., Plumptre, A., Hart, J. and Campos, L. (2002). Effects of war and civil strife on wildlife and wildlife habitats. *Conservation Biology*, 16(2), pp. 319–329.

Edelman, M., Weis, T., Baviskar, A., Borras, S., Holt-Giménez, E., Kandiyoti, D. and Wolford, W. (2014). Introduction: Critical perspectives on food sovereignty. *The Journal of Peasant Studies*, 41(6), pp. 911–931.

Etter, A., McAlpine, C., Pullar, D. and Possingham, H. (2005). Modeling the age of tropical moist forest fragments in heavily-cleared lowland landscapes of Colombia. *Forest Ecology and Management*, 208(1–3), pp. 249–260.

Etter, A., McAlpine, C., Wilson, K., Phinn, S. and Possingham, H. (2006). Regional patterns of agricultural land use and deforestation in Colombia. *Agriculture, Ecosystems and Environment*, 114(2–4), pp. 369–386.

FAO. (2016). *State of the World's Forests*. Rome: Food and Agriculture Organization of the United Nations.

Filotas, E., Parrott, L., Burton, P., Chazdon, R., Coates, K., Coll, L., Haeussler, S., Martin, K., Nocentini, S., Puettmann, K.J. and Putz, F.E. (2014). Viewing forests through the lens of complex systems science. *Ecosphere*, 5(1), pp. 1–23, art1.

Foley, J. (2005). Global consequences of land use. *Science*, 309(5734), pp. 570–574.

Foli, S., Reed, J., Clendenning, J., Petrokofsky, G., Padoch, C. and Sunderland, T. (2014). To what extent does the presence of forests and trees contribute to food production in humid and dry forest landscapes?: A systematic review protocol. *Environmental Evidence*, 3(1), p. 15.

Forero, J. (2002). The Colombian peasant economy 1990–2001, *Cuadernos de Tierra y Justicia*, no 2, ILSA, Bogotá.

Forero, J., Garay, L., Barberi, F., Ramírez, C., Suárez, M. and Gómez, R. (2013). The economic efficiency of large, medium and small Colombian agricultural producers. In: Garay Salamanca, L.J., Bailey, R., Forero, A., Barberi Gómez, F., Ramírez, G., Suárez, V., Myriam, D., Gómez, M., Castro Forero, Y., Zárate, Ã. and Manuel, J., eds, *Reflections on Rurality and Territory in Colombia: Problems and Current Challenges*, Bogotá: OXFAM.

Francis, C., Lieblein, G., Gliessman, S., Breland, T., Creamer, N., Harwood, R., Salomonsson, L., Helenius, J., Rickerl, D., Salvador, R., Wiedenhoeft, M., Simmons, S., Allen, P., Altieri, M., Flora, C. and Poincelot, R. (2003). Agroecology: The ecology of food systems. *Journal of Sustainable Agriculture*, 22(3), pp. 99–118.

Frost, P., Campbell, B., Medina, G. and Usongo, L. (2006). Landscape-scale approaches for integrated natural resource management in tropical forest landscapes. *Ecology and Society*, 11(2), p. 30.

Geist, H. and Lambin, E. (2002). Proximate causes and underlying driving forces of tropical deforestation. *BioScience*, 52(2), p. 143.

Gibbs, H., Ruesch, A., Achard, F., Clayton, M., Holmgren, P., Ramankutty, N. and Foley, J. (2010). Tropical forests were the primary sources of new agricultural land in the 1980s and 1990s. *Proceedings of the National Academy of Sciences*, 107(38), pp. 16732–16737.

Gliessman, S. (2015). *Agroecology: The Ecology of Sustainable Food Systems*. 3rd ed. Boca Raton: CRC Press.

Gonzalez, P. (2000). Coca, deforestation and food security in the Colombian Amazon region. *Unasylva*, 51(202), pp. 32–35.

Grau, H. and Aide, M. (2008). Globalization and land-use transitions in Latin America. *Ecology and Society*, 13(2), p. 16.

Hanson, T., Brooks, T.M., Da Fonseca, G.A., Hoffmann, M., Lamoreux, J.F., Machlis, G., Mittermeier, C.G., Mittermeier, R.A. and Pilgrim, J.D. (2009) Warfare in biodiversity hotspots. *Conservation Biology*, 23(3), pp. 578–587.

Holt-Gíménez, E. (2006*). Campesino a campes*ino. Oakland, CA: Food First Books.

Holt-Gíménez, E. and Altieri, M. (2012). Agroecology, food sovereignty and the new Green Revolution. *Journal of Sustainable Agriculture*, 37(1), pp. 90–102.

International Fund for Agricultural Development (IFAD). (2016). *Rural Development Report 2016: Fostering inclusive rural transformation*. Rome: Quintily.

Kremen, C., Iles, A. and Bacon, C. (2012). Diversified farming systems: An agroecological, systems-based alternative to modern industrial agriculture. *Ecology and Society*, 17(4).

Lambin, E., Geist, H. and Lepers, E. (2003). Dynamics of land-use and land-cover change in tropical regions. *Annual Review of Environmental Resources*, 28, pp. 205–241.

Lawry, S., Samii, C., Hall, R., Leopold, A., Hornby, D. and Mtero, F. (2016). The impact of land property rights interventions on investment and agricultural productivity in developing countries: A systematic review. *Journal of Development Effectiveness*, 9(1), pp. 61–81.

Lazos-Chavero, E., Zinda, J., Bennett-Curry, A., Balvanera, P., Bloomfield, G., Lindell, C. and Negra, C. (2016). Stakeholders and tropical reforestation: Challenges, trade-offs, and strategies in dynamic environments. *Biotropica*, 48(6), pp. 900–914.

Le Billon, P. (2001). The political ecology of war: Natural resources and armed conflicts. *Political Geography*, 20(5), pp. 561–584.

Liu, J., Dietz, T., Carpenter, S., Alberti, M., Folke, C., Moran, E., Pell, A.N., Deadman, P., Kratz, T., Lubchenco, J. and Ostrom, E. (2007). Complexity of coupled human and natural systems. *Science*, 317(5844), pp. 1513–1516.

Machlis, G. and Hanson, T. (2008). Warfare ecology. *BioScience*, 58(8), pp. 729–736.

Mansourian, S. (2016). Understanding the relationship between governance and forest landscape restoration. *Conservation and Society*, 14(3), p. 267.

Mbow, C., Smith, P., Skole, D., Duguma, L. and Bustamante, M. (2014). Achieving mitigation and adaptation to climate change through sustainable agroforestry practices in Africa. *Current Opinion in Environmental Sustainability*, 6, pp. 8–14.

McNeely, J. (2003). Biodiversity, war, and tropical forests. *Journal of Sustainable Forestry*, 16(3–4), pp. 1–20.

Messina, J. and Delamater, P. (2006). Defoliation and the war on drugs in Putumayo, Colombia. *International Journal of Remote Sensing*, 27(1), pp. 121–128.

Milburn, R. (2014). The roots to peace in the Democratic Republic of Congo: Conservation as a platform for green development. *International Affairs*, 90(4), pp. 871–887.

Minang, P., van Noordwijk, M., Freeman, O., Mbow, C., de Leeuw, J. and Catacutan, D. (2015). *Climate-Smart Landscapes: Multifunctionality in Practice*. Nairobi: World Agroforestry Centre (ICRAF).

Negret, P., Allan, J., Braczkowski, A., Maron, M. and Watson, J. (2017). Need for conservation planning in postconflict Colombia. *Conservation Biology*, 31(3), pp. 499–500.

Ordway, E. (2015). Political shifts and changing forests: Effects of armed conflict on forest conservation in Rwanda. *Global Ecology and Conservation*, 3, pp. 448–460.

Patel, R. (2009). Food sovereignty. *Journal of Peasant Studies*, 36(3), pp. 663–706.

Peluso, N. and Lund, C. (2011). New frontiers of land control: Introduction. *Journal of Peasant Studies*, 38(4), pp. 667–681.

Perfecto, I., Vandermeer, J. and Wright, A. (2009). *Nature's Matrix*. London: Earthscan.

Rahman, S., Jacobsen, J., Healey, J., Roshetko, J. and Sunderland, T. (2016). Finding alternatives to swidden agriculture: Does agroforestry improve livelihood options and reduce pressure on existing forest? *Agroforestry Systems*, 91(1), pp. 185–199.

Reed, J., Deakin, L. and Sunderland, T. (2015). What are 'Integrated Landscape Approaches' and how effectively have they been implemented in the tropics: A systematic map protocol. *Environmental Evidence*, 4(1), art 2.

Reed, J., Van Vianen, J., Deakin, E., Barlow, J. and Sunderland, T. (2016). Integrated landscape approaches to managing social and environmental issues in the tropics: Learning from the past to guide the future. *Global Change Biology*, 22(7), pp. 2540–2554.

Reyes Posada, A. (2009). *Guerreros y campesinos*. Bogotá (Colombia): Editorial Planeta Colombiana.

Rhemtulla, J. and Mladenoff, D. (2007). Why history matters in landscape ecology. *Landscape Ecology*, 22(S1), pp. 1–3.

Rincón-Ruiz, A. and Kallis, G. (2013). Caught in the middle, Colombia's war on drugs and its effects on forest and people. *Geoforum*, 46, pp. 60–78.

Rincón-Ruiz, A., Pascual, U. and Flantua, S. (2013a). Examining spatially varying relationships between coca crops and associated factors in Colombia, using geographically weight regression. *Applied Geography*, 37, pp. 23–33.

Rincón-Ruiz, A., Pascual, U. and Romero, M. (2013b). An exploratory spatial analysis of illegal coca cultivation in Colombia using local indicators of spatial association and socioecological variables. *Ecological Indicators*, 34, pp. 103–112.

Rosset, P., Patel, R. and Courville, M. (2006). *Promised Land*. Oakland, CA: Food First Books.

Sánchez-Cuervo, A. and Aide, T. (2013). Consequences of the armed conflict, forced human displacement, and land abandonment on forest cover change in Colombia: A multi-scaled analysis. *Ecosystems*, 16(6), pp. 1052–1070.

Sayer, J., Sunderland, T., Ghazoul, J., Pfund, J., Sheil, D., Meijaard, E., Venter, M., Boedhihartono, A.K., Day, M., Garcia, C. and van Oosten, C. (2013). Ten principles for a landscape approach to reconciling agriculture, conservation, and other competing land uses. *Proceedings of the National Academy of Sciences*, 110(21), pp. 8349–8356.

Schlüter, M., Mcallister, R., Arlinghaus, R., Bunnefeld, N., Eisenack, K., Hölker, F., Milner-Gulland, E.J., Müller, B., Nicholson, E., Quaas, M. and Stöven, M. (2012). New horizons for managing the environment: A review of coupled social-ecological systems modeling. *Natural Resource Modeling*, 25(1), pp. 219–272.

Scholes, R., Reyers, B., Biggs, R., Spierenburg, M. and Duriappah, A. (2013). Multi-scale and cross-scale assessments of social–ecological systems and their ecosystem services. *Current Opinion in Environmental Sustainability*, 5(1), pp. 16–25.

Souza, R., Mendes, I., Reis-Junior, F., Carvalho, F., Nogueira, M., Vasconcelos, A., Vicente, V.A. and Hungria, M. (2016). Shifts in taxonomic and functional microbial diversity with agriculture: How fragile is the Brazilian Cerrado? *BMC Microbiology*, 16(1), p. 42.

Tschakert, P., Coomes, O. and Potvin, C. (2007). Indigenous livelihoods, slash-and-burn agriculture, and carbon stocks in Eastern Panama. *Ecological Economics*, 60(4), pp. 807–820.

van Ausdal, S. (2009). Pasture, profit, and power: An environmental history of cattle ranching in Colombia, 1850–1950. *Geoforum*, 40(5), pp. 707–719.

van Oosten, C., Uzamukunda, A. and Runhaar, H. (2018). Strategies for achieving environmental policy integration at the landscape level. A framework illustrated with an analysis of landscape governance in Rwanda. *Environmental Science and Policy*, 83, pp. 63–70.

Villamor, G., van Noordwijk, M., Djanibekov, U., Chiong-Javier, M. and Catacutan, D. (2014). Gender differences in land-use decisions: Shaping multifunctional landscapes? *Current Opinion in Environmental Sustainability*, 6, pp. 128–133.

Vira, B., Wildburger, C. and Mansourian, S., eds. (2015). *Forests, Trees and Landscapes for Food Security and Nutrition. A Global Assessment Report*. IUFRO World Series. Vienna: International Union of Forest Research Organizations.

Visseren-Hamakers, I. (2015). Integrative environmental governance: Enhancing governance in the era of synergies. *Current Opinion in Environmental Sustainability*, 14, pp. 136–143.

Wunder, S., Angelsen, A. and Belcher, B. (2014). Forests, livelihoods, and conservation: Broadening the empirical base. *World Development*, 64, pp. S1–S11.

Young, O., Berkhout, F., Gallopin, G., Janssen, M., Ostrom, E. and van der Leeuw, S. (2006). The globalization of socio-ecological systems: An agenda for scientific research. *Global Environmental Change*, 16(3), pp. 304–316.

Part III

Integrated decision-making in forest landscape restoration

9 Stakeholders and forest landscape restoration

Who decides what to restore, why and where?

Stephanie Mansourian

Introduction

Stakeholders lie at the core of the forest landscape restoration (FLR) effort; they impact on restoration and are affected by it. Because FLR is intended to be at larger scales than much forest restoration undertaken to date, there are usually many more stakeholders present in the landscape, and many more power issues, unresolved conflicts and trade-offs to be negotiated (Sayer *et al.*, 2015; Reed *et al.*, 2016). For example, the forest service may promote some tree species that they are used to managing, while local communities in the landscape may prefer other species that provide them with a range of material and non-material benefits.

Truly understanding who the stakeholders are in the FLR process is important, not only to engage them, but also to understand and assess how they relate to the FLR process. For example, in Madagascar's Fandriana-Marolambo landscape, local communities who were key to the FLR effort were initially wary of engaging in the FLR project because of changes in legislation, which had first encouraged them to remove forests and was now punishing them for forest clearance and instead encouraging them to plant trees (Mansourian *et al.*, 2016).

Setting the boundary around 'who is in and who is out' in FLR is a political decision, with significant ramifications. For instance, in land use contexts, some distant but powerful stakeholder, such as a large private company, may be considered a 'priority stakeholder' by project managers or by the government, while less powerful but more forest-dependent communities may not.

The literature distinguishes diverse ways of categorizing stakeholders: they may be affected by the project, or affect it (Nutt and Backoff 1992; Mathur *et al.*, 2007); they may have the power to influence (Krott *et al.*, 2014); they may have a legitimate interest in the project (El-Gohary *et al.*, 2006); or they may be rightsholders or resource providers. 'Actors can be informed, consulted or co-produce knowledge and empowered' (Enengel *et al.*, 2012). Mitchell *et al.* (1997) propose that stakeholders can be identified as having one or more of three attributes: (1) the power to influence,

(2) legitimacy and (3) urgency of his/her claim on the object. Divergent definitions of stakeholders exist. For our purposes, stakeholders are defined as 'persons, groups, or organizations that must somehow be taken into account by leaders, managers, and front-line staff' (Bryson, 2004).

Understanding the role of stakeholders in forest loss and degradation is an important first step in determining actions for restoration. For example, conflict over land tenure may lead to land degradation. In Tunisia, for example, the 2011 revolution saw a significant degradation of national parks, as these were perceived to symbolize the oppressive regime. The expulsion of communities without legal tenure for the creation of national parks in several countries typifies these conflicts between environmental and social objectives (Geisler and de Sousa, 2000). In this case, understanding the role of the different actors in the conflict (typically, private land owners, communities and the government) will be critical to the success of restoration.

Stakeholder engagement is an essential element of restoration. Without effective stakeholder identification and understanding, groups may feel excluded, and the long-term viability and support for the restoration effort may be jeopardized (Reed *et al.*, 2009). Typical actions relating to stakeholders in restoring forested landscapes include identifying stakeholders, determining their relative dependence on the landscape/forest, identifying power relationships between them, determining their 'stake' in the endeavour, identifying their role in the restoration effort and defining acceptable means to engage them, determining whether they are winners or losers from the restoration intervention, and determining compensation measures should they be necessary (Adams *et al.*, 2016). Over the course of an FLR project or effort, the roles of stakeholders and their 'stakes' are likely to evolve and change, given the dynamic and long-term nature of the restoration effort (Mansourian, 2016). While categorizing groups of stakeholders may be useful, in practice it is important to not lose sight of the fact that individuals within groups of stakeholders may have very different motivations from the rest of the group (Blom *et al.*, 2010).

In this chapter, I explore, first, who are the stakeholders in FLR. Then I examine the way in which stakeholders are referred to in guidance documents on FLR and on large-scale forest restoration. The next section looks at stakeholder motivations, followed by an overview of the implications of restoration for stakeholders. Finally, I briefly discuss power dynamics in FLR.

Who are stakeholders in FLR?

At the level of individual landscapes, stakeholders will vary significantly, and generalizations can be dangerous. However, one can start to identify some categories of stakeholders based on professional disciplines (communities of practice) and also according to sectors of society (public, private or civil society).

Scientific disciplines and communities of practice

Three significant scientific disciplines and communities of practice can be identified as having a vested interest in forest landscape restoration: ecology, forestry and rural development.

Ecological community

Conservation biology, restoration ecology and landscape ecology may be seen as predominant influences on the development of FLR. Conservation biologists in the first half of the twentieth century focused almost exclusively on protected areas as the main tool to achieve their objectives of conserving a representative sample of species and their habitats. However, towards the late 1980s to early 1990s, they started realizing that there was a need to look beyond strictly protected areas, at the wider landscape matrix. As a result, they turned to landscape ecology to consider patterns and processes (Forman, 1995; Bennett, 1999; Morrison *et al.*, 2009). Equally, they gradually started incorporating – albeit reluctantly – restoration into their toolbox. There was (and still is) a reticence among conservation biologists to promote restoration as an approach, on the one hand, because of the implication that one can recreate a natural ecosystem, and on the other, because of the 'precautionary principle' (prevention being better than cure).

The scientific discipline of restoration ecology also evolved towards the last two decades of the twentieth century, with the formal establishment in 1988 of the Society for Ecological Restoration (SER), the guardian of the discipline. The SER defines ecological restoration as 'the process of assisting the recovery of an ecosystem that has been degraded, damaged or destroyed' (SER, 2004). While restoration ecology remains distinct from FLR, notably because it encompasses all ecosystems, focuses on historical reference ecosystems and does not explicitly seek to achieve social objectives, there have been changes in the scope and approach taken by many restoration ecologists. For example, a significant and recent break from traditional restoration ecology is the focus on 'novel ecosystems' that might be more adaptable and resilient to change than historically 'accurate' ones (e.g. Hobbs *et al.*, 2006; Choi, 2007).

Forestry community

In the context of forestry, forest restoration emerged as a form of management (Burton and Macdonald, 2011; Dey and Schweitzer, 2014). In many parts of the world, state forest agencies and the private sector have emphasized plantation forestry with single or few tree species, or natural forest management prioritizing timber production over other forest values, leading to simplified ecosystems and much criticism from the ecological

Figure 9.1 In rugged terrain such as the Swiss Alps, forests are important for avalanche control, as well as for their recreational value and for timber.
Source: photo © PJ Stephenson.

community (O'Farrell and Anderson, 2010; Boedhihartono and Sayer,
2012; Ciccarese *et al.*, 2012). As a result, for example, in Europe, many
forest species have been lost because of such management practices,
prompting calls for ecological restoration (Halme *et al.*, 2013). Nevertheless, today, in a growing number of countries (notably, the US and Switzerland), single-purpose forestry has been replaced with 'new forestry',
'multiple use' forests and 'close to nature' forestry, which emphasize the
multiple services forests provide – such as avalanche control, water protection, biodiversity conservation and so on (Gillis, 1990; Wiersum, 1995;
Kuchli and Blaser, 2005; Thompson *et al.*, 2011). At the same time, foresters have been giving growing attention to the wider landscape (Kohm and
Franklin, 1997). Forest research organizations, such as IUFRO (the International Union of Forest Research Organizations), CIFOR (the Center for
International Forestry Research) and universities, are demonstrating
increased interest in FLR, as evidenced by a rise in scientific publications
on FLR in forestry journals in recent years.

Rural development community

Because many areas that have potential for FLR (Minnemeyer *et al.*, 2011)
are rural and remote, there is a clear link between FLR and the rural

development community. Forest landscape restoration is being promoted by agencies such as the Food and Agriculture Organization (FAO) and the United Nations Convention to Combat Desertification (UNCCD) as a possible solution to food security and to land degradation (FAO and Global Mechanism of the UNCCD, 2015). The explicit inclusion of a 'human wellbeing' dimension to FLR, and the very fact that it seeks to improve the contribution of forests to wider landscapes, opens up the opportunity to achieve multiple objectives, including livelihood ones. Approaches such as community-based forest management have engaged local communities to manage forests, generally for multiple purposes, granting them some degree of responsibility while enabling them to obtain legal and recognized benefits from their management of the forests (such as collection of non-timber forest products) (Charnley and Poe, 2007). In the context of FLR, the establishment of community-managed fuelwood plantations has attempted to alleviate the impact on natural forests while empowering local communities and reducing their vulnerability. Communities have been documented to engage in effective forest restoration when given the rights and responsibility to manage forests (e.g. Poffenberger, 2006; Nagendra, 2007).

Sectoral actors

In addition to scientific communities, one can distinguish several sectoral actors in restoration.

Governments have become significant stakeholders in FLR, in particular further to the exponential growth of national commitments under the Bonn Challenge on FLR (Aronson and Alexander, 2013), which has sought to bring in various governments to publicly commit hectares of land to FLR. Two of the noteworthy national restoration initiatives that pre-dated the Bonn Challenge can be found in Brazil – the Atlantic Forest Restoration Pact – and in the US – the Collaborative Forest Landscape Restoration Program, which is the implementation tool of the FLR Act and is managed by the US Forest Service (since 2009, the FLR Act 'encourages the collaborative, science-based ecosystem restoration of priority forest landscapes' (Schultz *et al.*, 2012)).

Typically working within sectoral boundaries, government departments or ministries may not all collaborate towards the same objectives of restoration (see Chapter 2, this volume). Thus, for example, a country's agriculture department may provide subsidies to expand agriculture or tree crops, while its environment department might be emphasizing more diverse, indigenous and natural forest cover.

Private sector companies have a stake in FLR for numerous reasons. Increasingly, the insurance sector, for example, is seeing the benefits of promoting 'natural infrastructure' to reduce or mitigate the impacts of climate change (SwissRe, 2016). Also, because of the market for carbon credits starting with the Clean Development Mechanism of the United Nations Framework Convention on Climate Change (UNFCCC), and continuing under REDD+ (reducing emissions from deforestation and forest degradation, and fostering conservation, sustainable management of forests, and enhancement of forest carbon stocks), which promotes reforestation/afforestation as means of offsetting carbon emissions, many companies see the value in investing in restoration rather than reducing their carbon emissions.

Local communities – Those living in the landscape have the most to gain or lose from FLR. By virtue of their proximity to and dependence on the forest, rural communities derive multiple benefits from forests, including forest products and services (e.g. food, medicines, building materials, spiritual healing etc.). Restoration that addresses their needs may provide them with valuable goods and services. On the other hand, restoration may also generate conflict, exacerbate inequalities and negatively affect livelihoods (Adams *et al.*, 2016; Chapter 4, this volume). Furthermore, local, self-organized, community-level processes that could be labelled FLR, and that

Figure 9.2 Local women working in a community-run tree nursery in Rwanda.
Source: photo © S. Mansourian.

have not been reported as such by the mainstream literature, most certainly exist (e.g. Poffenberger, 2006) and deserve to be examined, as their knowledge of ecosystems may prove invaluable (Chapter 12, this volume).

Guidance documents on FLR and stakeholders

Engaging stakeholders appears in most restoration guidelines, although how to go about this, and how to go beyond a simple workshop bringing in a few representative stakeholders to inform them of a project, is rarely developed. Criticisms of superficial and token engagement of stakeholders abound (e.g. Cooke and Kothari, 2001). Yet, stakeholders living in the landscape are ultimately those who can ensure the sustainability of restoration.

A selection of guidance documents on restoration (large-scale restoration, not only FLR) reveals how some key organizations have proposed to engage stakeholders (see Table 9.1). Typical actions related to local stakeholders refer to 'integrating', 'facilitating', 'raising awareness', 'convincing', 'consulting' or 'collaborating'.

Different methods for stakeholder engagement exist (Sterling *et al.*, 2017). The reality and urgency of many restoration needs suggest that a suitable balance is required between overly simplified analyses of stakeholders and superficial stakeholder engagement versus rigorous detailed and comprehensive stakeholder mappings. Furthermore, shifts in stakeholder involvement and power are to be expected over time.

Leaders in forest landscape restoration: understanding stakeholder motivations

Many stakeholders have to come together to collaborate for restoration – even more so when dealing with landscape-scale restoration (Hodge and Adams, 2016). Stakeholders can engage in forest restoration for different reasons (see typology in Table 9.2): they may do so to comply with legislation, or because they receive subsidies; they may do so because of the religious, cultural or aesthetic values of a given species or a mix of species. Understanding motivations helps to build bridges across diverse stakeholders. A champion or leader will always be driving the FLR process, and understanding his/her motivations helps to better engage and collaborate with him/her.

Donor-led FLR – Donors may be governments, international organizations, or private foundations or individuals. Through their financial support, donors will seek to influence the direction of the project. Depending on their level of engagement, they may, for instance, determine the choice of location or the methods to be used. Inevitably, their expectations will influence the course of a project. For example, the German government has placed a significant amount of emphasis and funding on

Table 9.1 Guidance documents on restoration and stakeholders

Guidance document	Target and primary audience	Actions concerning stakeholders
International Tropical Timber Organization Guidelines for the Restoration, Management and Rehabilitation of Degraded and Secondary Tropical Forests (ITTO, 2002)	Target: forests Primary audience: public policymakers, and development and extension agencies; civil society, NGOs and private and communal extension agencies, forest practitioners, extension agents and others working at a site level; education, training and research institutions	'local communities and forest users should be fully integrated into the decision-making processes' 'clarify and legitimize equitable tenure, access, use and other customary rights […] among national and local stakeholders' 'facilitate access to information and organize training for all interested stakeholders […] on restoration' 'local communities and stakeholders […] participate in and share the responsibility for decision-making in […] restoration […] and rehabilitation'
The Society for Ecological Restoration International Primer on Ecological Restoration (SER, 2004)	Target: terrestrial, freshwater, coastal or marine ecosystems Primary audience: restoration practitioners	'collective decisions are more likely to be honoured and implemented'
An Attempt to Develop a Framework for Restoration Planning (Vallauri et al., 2005)	Target: forested landscapes Primary audience: project managers	'restorationists must […] raise awareness on the state of degradation in the landscape […] convince other stakeholders […] on the feasibility of forest restoration' 'restoration usually only works […] if it has support from a significant proportion of local stakeholders'

Document	Target / Primary audience	Relevant text
Ecological Restoration for Protected Areas: Principles, Guidelines and Best Practice (Keenleyside et al., 2012)	Target: protected areas Primary audience: protected area managers	'Step 1. [...] engage stakeholders' 'Consult and collaborate with all relevant partners and stakeholders [...] committed to the process'
Implementation of 2020 EU Biodiversity Strategy: Priorities for the restoration of ecosystems and their services in the EU (Lammerant et al., 2013)	Target: different ecosystems Primary audience: EU national governments Participatory consultation through working groups	'consultation with the stakeholders takes a central role' 'prioritization of ecosystem restoration should [...] be based on societal and political consensus, and [...] includes a form of stakeholder involvement' 'it is [...] essential that the affected people, organisations, and sectors (the stakeholders) are appropriately represented [...] in decisions on what to restore, where and when'
A Guide to the Restoration Opportunities Assessment Methodology (IUCN and WRI, 2014)	Target: forested landscapes Primary audience: decision-makers	'engaging key partners'
International Standards for the Practice of Ecological Restoration – Including Principles and Key Concepts (McDonald et al., 2016)	Target: terrestrial, freshwater, coastal or marine ecosystems Primary audience: practitioners, operational personnel, planners, managers, funders and regulators	'long-term relationship to place by local peoples (including Indigenous peoples) builds [...] knowledge of sites and ecosystems' 'restoration project managers should [...] engage with those who live or work within or near a restoration site, and those who have a stake in the area's biodiversity'

Source: adapted from Mansourian (2017).

Table 9.2 A typology to understand stakeholders' motivations for FLR

Motivation to restore		Actors				
Category	Reason	P	I	N	C	G
Economic	Corporate social responsibility	✓				
	Financial (or other) compensation	✓	✓	✓	✓	✓
	Reduce risk/liability	✓				✓
	Funding	✓	✓	✓	✓	✓
Cultural	Tradition		✓		✓	
	Spiritual value		✓	✓	✓	
	Heritage		✓	✓	✓	✓
	Identity		✓		✓	
Social	Recreation		✓		✓	✓
	Aesthetic value		✓		✓	
	Traditional knowledge/practices		✓		✓	
Ecosystem services	Erosion control	✓	✓	✓	✓	✓
	Food		✓	✓	✓	
	Pollination		✓	✓	✓	✓
	Water	✓	✓	✓	✓	✓
	Soil conservation	✓	✓	✓	✓	✓
	Microclimate regulation	✓	✓	✓	✓	✓
	Carbon sequestration	✓	✓	✓	✓	✓
Biodiversity conservation	Species habitat			✓	✓	✓
	Reproduction sites			✓	✓	
	Migration routes			✓	✓	
	Feeding areas for rare or endangered species			✓	✓	✓
Political	Commitment under global agreements					✓
	Policy/Legislation	✓		✓	✓	✓
	Elections					✓
	Impact on tenure rights		✓	✓	✓	

Key
P = private sector/company, I = individual/landholder, N = NGO,
C = community, G = government

obtaining commitments from other governments towards the Bonn Challenge on FLR. This has involved major investment in lobbying activities and preparation of reports, such as on the financial gains to be made through FLR (FAO and Global Mechanism of the UNCCD, 2015) or on the opportunities for achieving multiple global objectives.

Government-led FLR – Governments have the power to influence restoration both positively and negatively. For example, in cases where subsidies for agriculture outweigh any incentives to restore forests, there will more likely be a net loss of forests. This has been the case, for instance, in large parts of Latin America (e.g. Geist and Lambin, 2002). Equally, in cases where legislation and/or subsidies provide an incentive for restoration, there may be a greater attention paid to such practices. Brazil's Atlantic

Forest region has been widely publicized as an example, where the convergence of many actors, including public sector ones, has served to mobilize significant restoration action under the 'Atlantic Forest Restoration Pact'. Specific legislation, for instance in the state of São Paolo, has explicitly sought not only to outline the importance of restoration, but also to define the number and types of species to be used (Aronson *et al.*, 2011).

Company-led FLR – The private sector may engage in forest restoration to mitigate risk (from natural disasters, but also by protesters), to improve its image in the community, to comply with its corporate and social responsibility, or to respond to specific policies or norms (such as norms under certification schemes in the timber sector). For example, in Brazil's Bahia State, the pulp company Veracel planted about 400 hectares of native tree species every year, and by 2011 had restored a total of 4300 hectares of Atlantic rainforest (newgenerationplantations.org).

NGO-led FLR – Conservation non-governmental organizations (NGOs) have taken a lead in FLR, with the Worldwide Fund for Nature (WWF) and its multi-stakeholder partner International Union for Conservation of Nature (IUCN) defining the term in the year 2000. Restoration is viewed by such organizations as a complement to forest protection and management (Aldrich *et al.*, 2004). While protected areas remain the preferred tool of conservationists, in many areas, either within protected areas or around them, restoration is inevitable if biodiversity conservation objectives are to be met (Keenleyside *et al.*, 2012). Indeed, habitat loss remains the single highest cause of species extinction (Brooks *et al.*, 2002; Crouzeilles *et al.*, 2016). A meta-analysis of restoration projects led by Rey-Benayas *et al.* (2009) found that ecological restoration increased the provision of biodiversity by 44%.

Individual-led FLR – For various reasons, individuals may take the decision to restore part or all of their land. This may be because they have identified the importance of forests to their crops, or to improve the microclimate, or to protect their land, house or other infrastructure from landslides or floods. It may also be for biodiversity conservation or for cultural, religious or ethical reasons. In Scotland, for example, the significant loss of native forests, brought about to a large extent by English land owners and the centralized British government, has led to a link between restoration of native forests and nationalistic emotions (Colfer and Capistrano, 2005). Numerous land owners have thus engaged in restoration across Scotland (in part also spurred by public policies and subsidies).

Indigenous communities – There are many documented (and probably many more undocumented) cases of restoration that could 'qualify' as FLR by their scale and dual purpose, led by indigenous communities around the world. Thus, traditional pastoral systems that set aside land for it to naturally regenerate in the Atlas Mountains of Morocco (termed 'aqdals' (Auclair *et al.*, 2006)) or the religious forests created around Coptic churches in Ethiopia (Ruelle *et al.*, 2017) are important community-led efforts that may not have been planned as such at the scale of the entire landscape, but that ultimately add up to an important mosaic of forests across the landscape serving multiple purposes.

Such a typology may complement stakeholder mappings that help to identify stakeholders but without understanding their specific motivations for restoration. It may provide support for negotiating and reconciling trade-offs among stakeholders.

Roles of stakeholders in FLR

Not only will different stakeholders have different motivations to engage in an FLR process; they will also play different roles in the FLR process. The roles assumed relate to the stages in the FLR process but also to differing power dynamics. Table 9.3 suggests general roles that stakeholders may take in FLR. Stakeholders may also be categorized as winners or losers in FLR; they may be active or passive in the FLR process, or they may be proponents or recipients of an FLR project.

People take decisions, act, respond and engage in actions such as FLR based on their existing frame of reference. In turn, this frame of reference is influenced by a range of contextual factors, such as culture, politics, economics and so on. A contextual understanding of stakeholders serves to minimize generalizations and better understand what influences them. For example, in Canada's Cape Breton Highlands National Park, the Mi'qmak community, which was a key partner in the restoration of part of the park, has a different perception of geographic scale than that of Parks Canada. While the Mi'qmak perceive the wider landscape, shaped by the resources that they require, such as moose, Parks Canada focuses on the jurisdictional boundaries of the National Park (Mansourian *et al.*, forthcoming).

Implications of restoration for stakeholders

Restoring forest landscapes is lauded by many as providing benefits to stakeholders, particularly due to its ability to return or enhance ecosystem services such as water and soil conservation, pollination, carbon sequestration and so on. However, it may also imply costs, particularly opportunity costs, as the land may no longer be used for other purposes such as agriculture, housing or tourism. Restoration is associated with land improvements, yet for individual land owners, short-term financial gains may dictate other choices for their land, and trade-offs may be required (Bullock *et al.*, 2011). Some companies, on the other hand, may be willing to pay the price of forest restoration on part of their land because of the benefits perceived; for example, protection from flooding in coastal or riparian areas.

Table 9.3 Role of stakeholders in FLR

Role	Stakeholder group	Examples
Provides money/ invests	Company, NGO, government, landholder	Mining company supporting FLR as part of its corporate social responsibility (see, e.g., Parrotta and Knowles, 2001)
Sets policy framework for FLR	National government, intergovernmental organization	National government designing an FLR strategy (e.g. the US Congressional Act on Forest Landscape Restoration passed in 2009 – Schultz *et al.*, 2012)
Designs/plans project/intervention	NGO, community, company, landowner, academia, national forest service	Environmental NGO designing an FLR project to safeguard an endangered species (e.g. the Royal Society for the Protection of Birds co-purchasing Cousin Island in the Seychelles to restore habitat for the endemic Seychelles warbler – Komdeur and Pels, 2005)
Implements project/ intervention(s)	NGO, community, company, landowner, national forest service	Community engaged to remove invasive species as part of an FLR project (e.g. WWF engaging the public on special days to remove exotic species and plant indigenous ones in New Caledonia's dry forest – Mansourian and Vallauri, 2014)
Advises project/ intervention	NGO, community, company, government, academia, national forest service	Researchers providing technical advice on what to restore and where (within the landscape) in order to achieve intended objectives (e.g. support to local authorities and communities by Chiang Mai University's Forest Restoration Research Unit in Northern Thailand to restore Doi Suthep-Pui National Park – Elliott *et al.*, 2012)
Monitors progress of FLR project/ intervention	NGO, community, government, company, academia, national forest service	Local communities measuring progress with respect to forest restoration (e.g. local communities in Tanzania engaged in simple monitoring of forest restoration – Funder *et al.*, 2013)

Source: Mansourian (2017), reproduced with permission.

There have been documented cases (e.g. McElwee, 2009) of attempts at restoring forests (or at least planting trees) which have dispossessed communities of their land. There is a real risk that moves to restore large areas of forests may further displace rural communities (also see Chapter 3, this volume). More recently, discussions and debates over REDD+ and its implications for communities and land owners have also come to the fore (e.g. Phelps *et al.*, 2010; McLain *et al.*, 2017). Indeed, efforts to restore forested areas that require the secure long-term permanence of carbon sequestration (as per REDD+ requirements) may lead to the recentralization of forests and land.

Restoration may empower or, instead, weaken the claims of individual stakeholders. In Morocco, while the valuable argan fruit may fall under individual property rights, the trees themselves remain the property of the state, thereby reducing the incentive to restore large areas of argan forest (Biermayr-Jenzano *et al.*, 2014). In contrast, in Ghana, certificates attesting to the tenure of individual trees provide an incentive for restoration (Mansourian *et al.*, forthcoming). Challenges related to gender inequalities may also arise. For example, in many countries, such as Zimbabwe, women are reluctant to plant trees because of their insecurity of tenure, especially once they are widowed (e.g. Fortmann *et al.*, 1997).

Relationships among stakeholders and power dynamics

Imbalances of power have long plagued natural resource use where more powerful actors take decisions affecting less powerful ones. Power can be understood as a social relationship between two actors whereby one has the ability to affect or modify the behaviour of the other. In order to distinguish power from other means of modifying behaviour, Krott *et al.* (2014) identify three key elements of power: coercion, (dis)-incentives and dominant information, all of which confer upon one group the ability to influence the behaviour of the other.

When considering large-scale forest restoration or restoration in landscapes, power dynamics influence the choice of tree species, the location of trees and ultimately, the sustainability of the endeavour. This may be at the international scale (e.g. between a donor and a recipient country), at the national level (e.g. between public and private actors), at the landscape level (e.g. between different private actors) and even at the household level (e.g. between women and men). The REDD+ debate, and its associated risks to livelihoods because of the threat of more powerful actors (both national and international) appropriating land and forest, exemplifies these power imbalances (e.g. Phelps *et al.*, 2010). When restoring forests, one key decision relates to the choice of species, with individual species holding different values for each stakeholder. For example, some species may hold important medicinal and traditional value for some stakeholders, but may not be of interest to more powerful stakeholders, who may prefer to promote timber-producing species. Importantly, in terms of economics, the net result may appear positive, with a high financial value for the timber-producing species; however, this may fail to consider distributional aspects across stakeholders. As a result, less powerful actors may be much worse off because of restoration using species from which they do not benefit. 'Elite capture' in many conservation (and restoration) projects has been long documented (e.g. Nagendra, 2007). In a similar vein, the role of powerful international actors in determining restoration goals and objectives, and frequently 'setting the agenda' for large-scale interventions outside of their national boundaries, raises disturbing questions of legitimacy (Adams *et al.*, 2014).

Conclusion

Stakeholders are not static, and their relationship to the FLR process will evolve over time. The importance and relevance of FLR to their livelihoods will also change. Furthermore, while categorizing stakeholders provides a convenient way of analysing them and their relationship to restoration, each individual within a group may have vastly different objectives or constraints influencing their perception of, engagement with and acceptance of a process such as restoration. Restoring forest landscapes requires an understanding not only of who are the key stakeholders, but also of their relationship to each other and to the restoration process, motivations and underlying power dynamics. This information is necessary to ensure effective and fair negotiations when considering trade-offs among stakeholders and between people's needs and ecological priorities.

References

Adams, C., Rodrigues, S.T., Calmon, M. and Kumar, C. (2016) 'Impacts of large-scale forest restoration on socioeconomic status and local livelihoods: what we know and do not know', *Biotropica*, vol 48, no 6, pp. 731–744.

Adams, W.M., Aveling, R., Brockington, D., Dickson, B., Elliott, J., Hutton, J., Roe, D., Vira, B. and Wolmer, W. (2004) 'Biodiversity conservation and the eradication of poverty', *Science*, vol 306, no 5699, pp. 1146–1149.

Aldrich, M., Belokurov, A., Bowling, J., Dudley, N., Elliott, C., Higgins-Zogib, L., Hurd, J., Lacerda, L., Mansourian, S., McShane, T., Pollard, D., Sayer, J. and Schuyt, K. (2004) *Integrating Forest Protection, Management and Restoration at a Landscape Scale*, WWF International, Gland.

Aronson, J. and Alexander, S. (2013) 'Ecosystem restoration is now a global priority: time to roll up our sleeves', *Restoration Ecology*, vol 21, no 3, pp. 293–296.

Aronson, J., Brancalion, P.H., Durigan, G., Rodrigues, R.R., Engel, V.L., Tabarelli, M., Torezan, J., Gandolfi, S., de Melo, A.C., Kageyama, P.Y. and Marques, M. (2011) 'What role should government regulation play in ecological restoration? Ongoing debate in São Paulo State, Brazil', *Restoration Ecology*, vol 19, no 6, pp. 690–695.

Auclair, L., Bourbouze, A., Dominguez, P. and Genin, D. (2006) *Les agdals du Haut Atlas (Maroc) Biodiversité et gestion communautaire de l'accès aux ressources forestières et pastorales*, IRD, Marseille.

Bennett, A.F. (1999) *Linkages in the Landscape: the Role of Corridors and Connectivity in Wildlife Conservation*, IUCN, Gland.

Biermayr-Jenzano, P., Kassam, S.N. and Aw-Hassan, A. (2014) *Understanding Gender and Poverty Dimensions of High Value Agricultural Commodity Chains in the Souss-Masaa-Draa Region of South-western Morocco*, ICARDA working paper, mimeo, Amman.

Blom, B., Sunderland, T. and Murdiyarso, D. (2010) 'Getting REDD to work locally: lessons learned from integrated conservation and development projects', *Environmental Science & Policy*, vol 13, no 2, pp. 164–172.

Boedhihartono, A.K. and Sayer, J. (2012) 'Forest landscape restoration: restoring what and for whom?', in J. Stanturf, D. Lamb and P. Madsen (eds), *Forest Landscape Restoration: Integrating Natural and Social Sciences*, Springer, Dordrecht.

Brooks, T.M., Mittermeier, R.A., Mittermeier, C.G., Da Fonseca, G.A., Rylands, A.B., Konstant, W.R., Flick, P., Pilgrim, J., Oldfield, S., Magin, G. and Hilton-Taylor, C. (2002) 'Habitat loss and extinction in the hotspots of biodiversity', *Conservation Biology*, vol 16, no 4, pp. 909–923.

Bryson, J.M. (2004) 'What to do when stakeholders matter: stakeholder identification and analysis techniques', *Public Management Review*, vol 6, no 1, pp. 21–53.

Bullock, J.M., Aronson, J., Newton, A.C., Pywell, R.F. and Rey-Benayas, J.M. (2011) 'Restoration of ecosystem services and biodiversity: conflicts and opportunities', *Trends in Ecology & Evolution*, vol 26, no 10, pp. 541–549.

Burton, P.J. and Macdonald, S.E. (2011) 'The restorative imperative: challenges, objectives and approaches to restoring naturalness in forests', *Silva Fennica*, vol 45, no 5, pp. 843–863.

Charnley, S. and Poe, M.R. (2007) 'Community forestry in theory and practice: where are we now?' *Annual Review of Anthropology*, vol 36, pp. 301–336.

Choi, Y.D. (2007) 'Restoration ecology to the future: a call for new paradigm', *Restoration Ecology*, vol 15, no 2, pp. 351–353.

Ciccarese, L., Mattsson, A. and Pettenella, D. (2012) 'Ecosystem services from forest restoration: thinking ahead', *New Forests*, vol 43, no 5–6, pp. 543–560.

Colfer, C.J. and Capistrano, D. (eds), (2005) *The Politics of Decentralization: Forests, People and Power*, Earthscan, London.

Cooke, B. and Kothari, U. (2001) *Participation: The New Tyranny?* Zed Books, London.

Crouzeilles, R., Curran, M., Ferreira, M.S., Lindenmayer, D.B., Grelle, C.E. and Benayas, J.M.R. (2016) 'A global meta-analysis on the ecological drivers of forest restoration success', *Nature Communications*, vol 7, p. 11666.

Dey, D.C. and Schweitzer, C.J. (2014) 'Restoration for the future: endpoints, targets, and indicators of progress and success', *Journal of Sustainable Forestry*, vol 33, pp. S43–S65.

El-Gohary, N.M., Osman, H. and El-Diraby, T.E. (2006) 'Stakeholder management for public private partnerships', *International Journal of Project Management*, vol 24, pp. 595–604.

Elliott, S., Kuaraksa, C., Tunjai, P., Toktang, T., Boonsai, K., Sangkum, S., Suwannaratana, S. and Blakesley, D. (2012) 'Integrating scientific research with community needs to restore a forest landscape in northern Thailand: a case study of Ban Mae Sa Mai', in J. Stanturf, D. Lamb and P. Madsen (eds), *Forest Landscape Restoration: Integrating Natural and Social Sciences*, Springer, Dordrecht.

Enengel, B., Muhar, A., Penker, M., Freyer, B., Drlik, S. and Ritter, F. (2012) 'Co-production of knowledge in transdisciplinary doctoral theses on landscape development – an analysis of actor roles and knowledge types in different research phases', *Landscape and Urban Planning*, vol 105, no 1, pp. 106–117.

FAO and Global Mechanism of the UNCCD. (2015) *Sustainable Financing for Forest and Landscape Restoration*, FAO and UNCCD, Rome and Bonn.

Forman, R.T.T. (1995) 'Some general principles of landscape and regional ecology', *Landscape Ecology*, vol 10, no 3, pp. 133–142.

Fortmann, L., Antinori, C. and Nabane, N. (1997) 'Fruits of their labors: gender, property rights, and tree planting in two Zimbabwe villages', *Rural Sociology*, vol 62, no 3, pp. 295.

Funder, M., Danielsen, F., Ngaga, Y., Nielsen, M.R. and Poulsen, M.K. (2013) 'Reshaping conservation: the social dynamics of participatory monitoring in Tanzania's community-managed forests', *Conservation and Society*, vol 11, no 3, p. 218.

Geisler, C. and De Sousa, R. (2001) 'From refuge to refugee: the African case', *Public Administration and Development*, vol 21, no 2, pp. 159–170.

Geist, H.J. and Lambin, E.F. (2002) 'Proximate causes and underlying driving forces of tropical deforestation: tropical forests are disappearing as the result of many pressures, both local and regional, acting in various combinations in different geographical locations', *BioScience*, vol 52, no 2, pp. 143–150.

Gillis, A.M. (1990) 'The new forestry', *BioScience*, vol 40, no 8, pp. 558–562.

Halme, P., Allen, K.A., Auni š, A., Bradshaw, R.H., Br melis, G., ada, V., Clear, J.L., Eriksson, A.M., Hannon, G., Hyvärinen, E. and Ikauniece, S. (2013) 'Challenges of ecological restoration: lessons from forests in northern Europe', *Biological Conservation*, vol 167, pp. 248–256.

Hobbs, R.J., Arico, S., Aronson, J., Baron, J.S., Bridgewater, P., Cramer, V.A., Epstein, P.R., Ewel, J.J., Klink, C.A., Lugo, A.E. and Norton, D. (2006) 'Novel ecosystems: theoretical and management aspects of the new ecological world order', *Global Ecology and Biogeography*, vol 15, no 1, pp. 1–7.

Hodge, I. and Adams, W.M. (2016) 'Short-term projects versus adaptive governance: conflicting demands in the management of ecological restoration', *Land*, vol 5, p. 39.

ITTO. (2002) *International Tropical Timber Organization Guidelines for the Restoration, Management and Rehabilitation of Degraded and Secondary Tropical Forests*, ITTO, Yokohama.

IUCN and WRI. (2014) *A Guide to the Restoration Opportunities Assessment Methodology (ROAM)*, IUCN and WRI, Gland and Washington DC.

Keenleyside, K., Dudley, N., Cairns, S., Hall, C. and Stolton, S. (2012) *Ecological Restoration for Protected Areas: Principles, Guidelines and Best Practice*, IUCN, Gland.

Kohm, K.A. and Franklin, J.F. (eds) (1997) *Creating a Forestry for the 21st Century: The Science of Ecosystem Management*, Island Press, Washington DC.

Komdeur, J. and Pels, M.D. (2005) 'Rescue of the Seychelles warbler on Cousin Island, Seychelles: the role of habitat restoration', *Biological Conservation*, vol 124, no 1, pp. 15–26.

Krott, M., Bader, A., Schusser, C., Devkota, R., Maryudi, A., Giessen, L. and Aurenhammer, H. (2014) 'Actor-centred power: the driving force in decentralised community based forest governance', *Forest Policy and Economics*, vol 49, pp. 34–42.

Kuchli, C. and Blaser, J. (2005) 'Forests and decentralization in Switzerland: a sampling', in C.J.P. Colfer and D. Capistrano (eds), *Politics of Decentralization: Forests, People and Power*, pp. 152–165, Earthscan, Abingdon.

Lammerant, J., Peters, R., Snethlage, M., Delbaere, B., Dickie, I. and Whiteley, G. (2013) *Implementation of 2020 EU Biodiversity Strategy: Priorities for the Restoration of Ecosystems and their Services in the EU*. Report to the European Commission. ARCADIS (in cooperation with ECNC and Eftec).

Mansourian, S. (2016) 'Understanding the relationship between governance and forest landscape restoration', *Conservation and Society*, vol 14, no 3, pp. 267.

Mansourian, S. (2017) 'Governance and Forest Landscape Restoration: a framework to support decision-making', *Journal for Nature Conservation*, vol 37, pp. 21–30.

Mansourian, S. and Vallauri, D. (2014) 'Restoring forest landscapes: important lessons learnt', *Environmental Management*, vol 53, no 2, pp. 241–251.

Mansourian, S., Razafimahatratra, A., Ranjatson, P. and Rambeloarisoa, G. (2016) 'Novel governance for forest landscape restoration in Fandriana-Marolambo, Madagascar', *World Development Perspectives*, vol 3, pp. 28–31.

Mansourian, S. *et al.* (forthcoming) Identifying governance problems and negotiating solutions for forest landscape restoration in New Caledonia, Canada, and Ghana.

Mathur, V.N., Price, A.D., Austin, S.A. and Moobela, C. (2007) 'Defining, identifying and mapping stakeholders in the assessment of urban sustainability', in M. Horner, C. Hardcastle, A. Price and J. Bebbington (eds), *Proceedings: SUE-MoT Conference 2007: International Conference on Whole Life Sustainability and its Assessment*, Glasgow, Scotland, 27–29 June 2007.

McDonald, T., Gann, G.D., Jonson, J. and Dixon, K.W. (2016) *International Standards for the Practice of Ecological Restoration – Including Principles and Key Concepts*. Society for Ecological Restoration, Washington, DC.

McElwee, P. (2009) 'Reforesting "bare hills" in Vietnam: social and environmental consequences of the 5 million hectare reforestation program', *Ambio*, vol 38, no 6, pp. 325–333.

McLain, R., Guariguata, M.R. and Lawry, S. (2017) *Implementing Forest Landscape Restoration Initiatives: Tenure, Governance, and Equity Considerations*, GIZ and CIFOR, Bonn and Bogor.

Minnemeyer, S., Laestadius, L. and Sizer, N. (2011) *A World of Opportunity*, World Resources Institute, Washington DC.

Mitchell, R.K., Agle, B.R. and Wood, D.J. (1997) 'Toward a theory of stakeholder identification and salience: defining the principle of who and what really counts', *Academy of Management Review*, vol 22, no 4, pp. 853–886.

Morrison, J., Loucks, C., Long, B. and Wikramanayake, E. (2009) 'Landscape-scale spatial planning at WWF: a variety of approaches', *Oryx*, vol 43, no 04, pp. 499–507.

Nagendra, H. (2007) 'Drivers of reforestation in human-dominated forests', *Proceedings of the National Academy of Sciences*, vol 104, no 39, pp. 15218–15223.

Nutt, P. and Backoff, R. (1992) *Strategic Management of Public and Third Sector Organizations: A Handbook for Leaders*, Jossey-Bass, San Francisco, CA.

O'Farrell, P.J. and Anderson, P.M. (2010) 'Sustainable multifunctional landscapes: a review to implementation', *Current Opinion in Environmental Sustainability*, vol 2, no 1, pp. 59–65.

Parrotta, J.A. and Knowles, O.H. (2001) 'Restoring tropical forests on lands mined for bauxite: examples from the Brazilian Amazon', *Ecological Engineering*, vol 17, no 2–3, pp. 219–239.

Phelps, J., Webb, E.L. and Agrawal, A. (2010) 'Does REDD+ threaten to recentralize forest governance?' *Science*, vol 328, pp. 312–313.

Poffenberger, M. (2006) 'People in the forest: community forestry experiences from Southeast Asia', *International Journal of Environment and Sustainable Development*, vol 5, no 1, pp. 57–69.

Reed, J., Van Vianen, J., Deakin, E.L., Barlow, J. and Sunderland, T. (2016) 'Integrated landscape approaches to managing social and environmental issues in the tropics: learning from the past to guide the future', *Global Change Biology*, vol 22, no 7, pp. 2540–2554.

Reed, M.S., Graves, A., Dandy, N., Posthumus, H., Hubacek, K., Morris, J., Prell, C., Quinn, C.H. and Stringer, L.C. (2009) 'Who's in and why? A typology of stakeholder analysis methods for natural resource management', *Journal of Environmental Management*, vol 90, no 5, pp. 1933–1949.

Rey-Benayas, J.M.R., Newton, A.C., Diaz, A. and Bullock, J.M. (2009) 'Enhancement of biodiversity and ecosystem services by ecological restoration: a meta-analysis', *Science*, vol 325, no 5944, pp. 1121–1124.

Ruelle, M.L., Kassam, K.A. and Asfaw, Z. (2017) 'Human ecology of sacred space: church forests in the highlands of northwestern Ethiopia', *Environmental Conservation* pp. 1–10. https://doi.org/10.1017/S0376892917000534.

Sayer, J., Margules, C., Boedhihartono, A.K. and Dale, A. (2015) 'Landscape approaches; what are the pre-conditions for success?' *Sustainability Science*, vol 10, no 2, pp. 345.

Schultz, C.A., Jedd, T. and Beam, R.D. (2012) 'The Collaborative Forest Landscape Restoration Program: a history and overview of the first projects', *Journal of Forestry*, vol 110, no 7, pp. 381–391.

SER (Society for Ecological Restoration) International Science & Policy Working Group. (2004) *The SER International Primer on Ecological Restoration*, www.ser.org and Society for Ecological Restoration International, Tucson.

Sterling, E.J., Betley, E., Sigouin, A., Gomez, A., Toomey, A., Cullman, G., Malone, C., Pekor, A., Arengo, F., Blair, M. and Filardi, C. (2017) 'Assessing the evidence for stakeholder engagement in biodiversity conservation', *Biological Conservation*, vol 209, pp. 159–171.

SwissRe. (2016) *Natural Catastrophes and Man-Made Disasters in 2015: Asia Suffers Substantial Losses*, SwissRe, Zurich.

Thompson, I.D., Okabe, K., Tylianakis, J.M., Kumar, P., Brockerhoff, E.G., Schellhorn, N.A., Parrotta, J.A. and Nasi, R. (2011) 'Forest biodiversity and the delivery of ecosystem goods and services: translating science into policy', *BioScience*, vol 61, no 12, pp. 972–981.

Vallauri, D., Aronson, J. and Dudley, N. (2005) 'An attempt to develop a framework for restoration planning', in S. Mansourian, D. Vallauri and N. Dudley (eds), *Forest Restoration in Landscapes: Beyond Planting Trees*, Springer, New York.

Wiersum, K.F. (1995) '200 years of sustainability in forestry: lessons from history', *Environmental Management*, vol 19, no 3, pp. 321–329.

10 Tenure, property rights and forest landscape restoration

Wil de Jong, Marieke van der Zon,
Andrea Flores Urushima, Yeo-Chang Youn,
Jinlong Liu and Ning Li

Introduction

Restoring forest cover is high on the agenda of national governments and organizations whose mandate relates to sustainable development or environmental conservation. A number of international initiatives include commitments to meet forest restoration targets, such as the Bonn Challenge on forest landscape restoration, the New York Declaration on Forests and the UN Strategic Plan for Forests (see Chapter 1 and de Jong (2018) for a brief review of these initiatives).

There is much interest among those who support or promote forest landscape restoration in understanding how these ambitious targets can be achieved, or more generally, how large-scale forest restoration can be implemented successfully. An important proportion of the literature that has addressed forest restoration has focused on more technical aspects of bringing back forest where it has disappeared, or restoring the forest condition of degraded forests. In the early 1990s, Lamb (1994) pointed out the need to address non-technical aspects that influence forest restoration. They include governance and institutional factors as well as economic, social and cultural factors. Tenure is widely considered to be an important governance or institutional factor that influences forest landscape restoration (Mansourian, 2017).

The importance of tenure and property rights in the context of land use and forests has been recognized for many years. Alchian and Demsetz (1973) defined tenure or property rights as 'socially recognized rights to action'. The authors postulate the property rights paradigm, which argues that economic actors are more likely to make investments, in terms of financial or other resources, labour and land, when they have expectations that they may capture the benefits that will result from these investments. Secure or legal tenure over the asset in which investments are made and which generates benefits will increase the expectation that benefits can be captured, and thus encourage economic investment. A distinction is to be made between tenure and property rights. Tenure refers to the 'terms on which something is held: the rights and obligations of the holder', while property rights refer to the specific group or 'bundle' of rights that are associated with the asset held in tenure (Bruce, 1998).

This chapter aims to analyse tenure and its role in forest landscape restoration. It will try to capture the relevance and importance of tenure considerations when forest landscape restoration is being pursued or supported. The chapter is entirely based on review of existing literature and documentation. The material available for review provides a complex and not unequivocal picture of how tenure relates to forest landscape restoration. In order to adequately represent the evidence, the chapter will first briefly reflect on the two key concepts that are important here, forest landscape restoration and tenure, and on the ongoing 'tenure'–'forest landscape restoration' debate, by reviewing related narratives, or what Woods (2010) refers to as the 'tenure security discourse'. We will then present the empirical evidence of tenure and forest landscape restoration, considering four cases of forest landscape restoration in Asia and highlighting the role of tenure in these efforts. This will be followed by a broader discussion of formalization of tenure and forest landscape restoration. We will then explore findings of general relevance, followed by conclusions.

Definitions, narratives and their implications for forest landscape restoration

For about a decade, the concept of forest landscapes has become widely adopted, with forest landscape restoration becoming a major focus in research and policy circles. The latter is defined by Mansourian *et al.* (2005) as 'a planned process that aims to regain ecological integrity and enhance human well-being in deforested or degraded forest landscapes'. The International Union for Conservation of Nature (IUCN) website expands that definition to:

> the ongoing process of regaining ecological functionality and enhancing human well-being across deforested or degraded forest landscapes. Forest landscape restoration is more than just planting trees – it is restoring a whole landscape 'forward' to meet present and future needs and to offer multiple benefits and land uses over time.[1]

The same webpage, furthermore lists 'different processes such as: new tree plantings, managed natural regeneration, agroforestry, or improved land management to accommodate a mosaic of land uses, including agriculture, protected wildlife reserves, managed plantations, riverside plantings and more'. Forest landscape restoration aims to improve forest area and forest quality in landscapes, but without necessarily covering the whole landscape with trees.

A definition of what is forest landscape restoration matters when considering the relevance of tenure. When debating forest restoration, tenure considerations can be limited to the specific location on which forest is expected to be restored, which, as suggested by Chokkalingam *et al.* (2005), occurs on formerly forested grasslands, brushlands, scrublands or barren areas. Probably agricultural lands can be added to that list, as in a significant number of countries in the world net forest cover has increased, and much of this forest has been restored on previous agricultural lands (de Jong *et al.*, 2017a). A discussion of tenure and forest restoration could thus consider the tenure condition or regime of the particular piece of land within the landscape that is being considered for forest restoration, and the influence of this tenure regime on the likelihood, feasibility and likely success of forest restoration efforts, not only in this particular location but also for the landscape as a whole. However, as additional specifications are added to the definition of forest landscape restoration, the questions also become more complex, including whether tenure will affect *how* forest restoration may be implemented. The latter influences the extent to which forest restoration may contribute to ecological objectives and, ultimately, also how the tenure condition may influence human wellbeing outcomes of forest restoration in a particular landscape. In such a scenario, diverse and at times conflicting tenure regimes and property rights may complicate matters. Within a landscape, as opposed to a forest patch, there are more likely to be several land owners, public and/or private. Conflict or disagreements on tenure rights are also more likely. In turn, this may create challenges for restoring forests within the landscape. While tenure security is not essential to engage in restoration, tenure insecurity has been cited as a disincentive to invest in restoration (Fortmann and Bruce, 1991; Cotula and Mayers, 2009). Equally, forest restoration may in turn affect tenure, with, for example, cases where returning trees to land may entitle farmers to legal tenure to the land, such as in Madagascar (Mansourian *et al.*, 2016).

Furthermore, while tenure and property rights tend to be associated with land, they are also applicable to individual trees, the fruit and products from the trees, as well as the services (e.g. carbon sequestration) provided by the trees. Thus, different stakeholders may hold different property rights over each of these components, adding further complexity to the restoration process. For example, in Morocco, all argan trees are owned by the government, irrespective of whose land they are grown on (Biermayr-Jenzano *et al.*, 2014). Given the enthusiasm for carbon sequestration projects through tree planting, it is important to note, for example, that the rightsholder to the carbon may be very different from the rightsholder to the forest, land or trees (Sunderlin *et al.*, 2009).

Tenure is associated with a series of other terms, including tenure rights, legal tenure, secure tenure and tenure status. A more recent conceptualization is suggested by von Benda-Beckmann *et al.* (2006), who define (land) tenure as 'the social relations and institutions governing access to and use of land and resources'. While some definitions try to conceptualize tenure, other assessments make a distinction between some kind of formalized status of an asset and subsequent rights that are derived from that status (Payne, 2004). In that conceptualization, the tenure status refers to 'the mode by which land or property is owned or held'. This can be according to a range of categories or arrangements, including legal, customary, semi-legal, informal or religious, while within each of these categories, multiple variations are common. From each of the tenure statuses, property rights are derived. In the case of land, the latter can be characterized as 'what one is permitted to do with the land' (cf. Payne, 2004). Schlager and Ostrom (1992) identified five individual property rights: (1) the right of access, (2) the right of withdrawal, (3) the right of alienation, (4) the right of management and (5) the right of exclusion. More recently, building on their work, Galik and Jagger (2015) have added the right of alteration, which is particularly relevant to restoration, as it includes forest removal or its converse, restoration or tree planting.

Legal or statutory tenure

Within the tenure status, legal or statutory ownership is relevant, as, at least in theory, it commands state-institutional protection. Legal or statutory tenure is a formalized recognition of ownership. It is ownership that has been granted to a legal entity within a particular legal regime. The latter includes laws that recognize the tenure status and define which property rights are derived, and withheld, from ownership, and it includes regulations that stipulate the process that needs to be followed if legal tenure is desired or if it is to be transferred. The process to grant legal tenure over land is prescribed in common law, and it is a legal-administrative process for which the procedures are generally well defined. In addition, legal tenure is, at least in theory, enforceable in court or other relevant public institutions. Common law also stipulates what property rights the tenure holders enjoy and what obligations are implied by the holding of legal tenure over land (i.e. payment of taxes, among others). Calls for improving tenure or property rights in what Woods (2010) refers to as the tenure security discourse most commonly refer to the progressive granting of legal tenure.

While legal or statutory tenure generally provides clearer and more secure rights, legal tenure is not the only formal option to grant rights, as the examples in the following section clearly demonstrate. It is quite common for rights to be granted when forest restoration is being pursued, without legal tenure. Even where common-law legal tenure is being granted over forestland, significant rights may be withheld. Common law, for instance, may allow public authorities to make demands that affect the rights that the tenure holder enjoys. This is the case for Canada (Luckert, 1991; Haley and Luckert, 1992; Luckert *et al.*, 2011) and China (Liu, 2001; Sun *et al.*, 2017), for example, but it is common wherever legal tenure is granted over urban or rural spaces (Payne, 2004). New policies may be devised by government forest agencies, which change the rights and obligations of tenure holders even though the tenure status remains equal. In practice, property rights held by a legal tenure holder are thus highly diverse and context specific, and they can change over time, depending on public administration policies or changes in the law. Although legal tenure is widely believed to be stronger, it may in practice give fewer rights than less formal arrangements (Payne, 2004). This is the case in areas where the government is largely absent and does not give any practical protection, or when formal rights are highly restrictive and limit the use of resources.

The observed ambiguity between legal status and actual property rights persists in many locations in the world. Communities of households may hold legal tenure, but the actual rights that are held of land or resources are derived from informal or customary arrangements. Where communities hold collective land titles, the communal land is often subdivided based on informal or customary arrangements (e.g. Cano *et al.*, 2014; Cronkleton and Larson, 2015). While the arrangements and rights held are generally clear to the actors involved, they seem complex and are difficult to understand for outsiders, not least because arrangements may differ between and even within forest landscapes (Cano *et al.*, 2014).

Tenure and forest restoration narratives

The practical reality in locations where much of contemporary forest landscape restoration is at least considered to be necessary is that land use and tenure status are profoundly out of sync. In forest landscapes, it is common that land use which would require legal tenure – either individual or communal land titles – does not benefit from legal status. Ownership in most forest landscapes includes legal, customary or other informal tenure. Quite often, where communities reside and both agriculture and forest management are part of livelihood strategies, the land remains classified as state forestland. Not uncommonly, the state may grant the right to use such land under concessions for timber exploitation or estate crop production. It is also still common that individual households or communities have no legal tenure or other formally recognized ownership, or that they have these only over parts of the land and natural resources on which they depend to meet their livelihood needs.

The relevance of the tenure status or the property rights that forest users derive from land and natural resources is widely discussed in the literature. A commonly used approach to understanding the importance of particular concepts or ideas in contemporary debates (including in debates related to environmental issues) is to analyse the prevalent narratives or discourses related to them (e.g. Leipold, 2014; de Jong *et al.*, 2017b). These narratives or discourses are highly relevant, as they shape not only popular but also specialized perceptions of the concept and topic. These perceptions often can be highly influential in, for instance, policy decisions or other important decisions that are being made by governments, their agencies, national and international civil society organizations, development cooperation and others.

In the literature on forest conservation, forest and development, and forest restoration, a prevailing discourse on forest and tenure exists (Woods, 2010). This discourse holds that individuals or groups need formal ownership over forestland and forests to allow them to derive benefits, and that formal ownership will expand efforts to preserve those forests. The benefits can be economic, but also cultural, spiritual and even aesthetic. Narratives in the discourse restate the argument that legal or statutory ownership will motivate rightsholders to intensify their labour, management and investment efforts (Feder and Feeny, 1991; Besley, 1995), whereas without such formalized ownership they may be much less willing to do so.

Since the 1990s, influential international organizations such as the Rights and Resources Initiative (RRI), the Food and Agriculture Organization (FAO) and Oxfam, have actively promoted improving legal land ownership to promote economic growth and poverty reduction in developing countries (Woods, 2010). When applied to forests and forest-dependent communities, the discourse implies that legal ownership or other legal rights over forest land and resources are a prerequisite for achieving the dual objective of improving livelihoods by mobilizing forest benefits and assuring that capturing forest benefits leads to sustainable forest management and forest conservation outcomes. For example, in Nepal, Nagendra (2007) found that the likelihood of engagement in reforestation activities was in part determined by the tenure regime. The tenure security discourse is also prominently present in the forest landscape restoration literature. Lamb *et al.* (2005), for instance, argue that because of insecure land and tree tenure, rural people are unwilling to invest in reforestation, because they may derive little benefit from this, and that providing secure tenure might make reforestation more attractive to land owners. Several articles in a 2015 special issue of *Unasylva* on 'Forest and landscape restoration' make the same or similar points (e.g. Appanah *et al.*, 2015; Berrahmouni *et al.*, 2015; Laestadius *et al.*, 2015; Sabogal *et al.*, 2015). While it is beyond the scope of this chapter to fully investigate the tenure security discourse in the forest restoration literature, it is relevant to point out how persistent this discourse has turned out to be. Two examples demonstrate this. For instance, Mansourian *et al.* (2016) observe that improving tenure and rights is one of the options to stimulate forest landscape restoration, while Wilson and Cagalanan (2016) observe that land tenure is a key governance mechanism in forest landscape restoration, and that decentralization that improves access, exclusion rights and decision-making rights results in better outcomes.

Criticism of the property rights discourse

From the analysis of tenure and forest restoration narratives, it emerges that there is a widely shared assumption that a positive causal linkage exists between legal tenure, or other secure property rights arrangements over agricultural and forest lands, benefit streams that can be derived from these lands, and the likelihood that property holders will make management investments related to their lands (e.g. Lawry, 2015). The causal relationship, however, between legal tenure and economic development has been questioned mostly by scholars focusing on developing countries (Cronkleton and Larson, 2015). Scholars who challenge the tenure security discourse generally agree that tenure security is a precondition for rural development or forest conservation and restoration objectives. However, they consider the prevalent narrative that dominates the tenure security discourse to be somewhat simplistic (i.e. Cousins, 2005; Chang, 2011), as it focuses solely on legal tenure (i.e. a government-issued land title; Cronkleton and Larson, 2015). Critics, instead, point out limitations of legal tenure and argue that property rights security can be obtained from many sources that enjoy social legitimacy, and that, as Payne (2004) has argued, non-legal or formal types of tenure arrangement may result in more secure rights.

Secure property rights are often not derived from the legal tenure status, but instead, are based on residents' perceptions of past and present government policy (Payne, 2004). If the government, for example, lacks a policy to evict squatters from titled land, and this illegal squatting happens frequently, tenure security is low even though one holds legal tenure. Instead, where governments help people to secure tenure through practical activities instead of legalistic ones, people feel they have secure tenure and are willing to invest. Payne recommends focusing on improving reasonably secure tenure arrangements instead of focusing on providing the most formal titles. In the context of rural development and access to loans in Peru, it has been argued 'that government land titling programmes should not be automatically preferred over utilizing the existing local institutions' (Kerekes and Williamson, 2010). Rather, securing of property rights can in many locations be better achieved through informal private mechanisms, such as customary rules and enforcement mechanisms (Kerekes and Williamson, 2010; Cronkleton and Larson, 2015).

Empirical evidence of tenure arrangements and forest landscape restoration

In the following discussion, we focus on three cases in Asia in which land owners were given some kind of legal tenure as well as specific rights, and evaluate whether this, in turn, has had a positive impact on forest landscape restoration. The empirical evidence available does not provide an unequivocal picture of whether changes in tenure arrangements can have a positive impact on forest landscape restoration outcomes. This is because the majority of studies exploring factors that impact forest landscape restoration outcomes often have not been able (or did not attempt) to separate single factors, but commonly identify a number of factors that contribute to successful forest restoration.

A way to address similar epistemological challenges was explored by Katila *et al.* (2014), who reviewed a total of 27 case studies to identify linkages between influencing factors and forest management outcomes. The cases explored the narrative evidence of factors that hypothetically influence sustainable forest management, by analysing the frequency of these factors in relation to forest management outcomes. Based on their analysis, several key factors that appeared to be correlated with successful forest management outcomes could be detected. The analysis suggested causality between single factors and factors in combination and successful forest management outcomes. The terms used for expressing the linkages were 'prerequisite conditions' and 'prerequisite factor synergies' (Katila *et al.*, 2014). In the following, we discuss cases in which tenure and property rights adjustments were made in an effort to create the prerequisite conditions for forest landscape restoration, and which reportedly had positive forest landscape restoration outcomes.

Guangdong, China

A benchmark set of six country studies that reviewed forest restoration and its drivers was completed by the Center for International Forestry Research (CIFOR) during the early 2000s (Chokkalingam *et al.*, 2005). Based on the studies, the authors concluded that institutional support and arrangements, including clear and undisputed land tenure status, are indispensable for sustained forest restoration. Complementary key factors were also identified, including well-functioning local organizations and participation in rehabilitation efforts, as well as adequate considerations of local needs (Chokkalingam *et al.*, 2005).

Among these CIFOR studies, one specific example that is suitable for examining the linkage between land tenure, related rights and forest landscape restoration is from Guangdong, China. Since the 1950s, Guangdong has experienced similar land tenure reforms as the rest of China (Liu, 2001). Forest cover in Guangdong increased from 27% to 57% between 1985 and 2003 (Chokkalingam *et al.*, 2006). This forest cover change is said to have occurred as a result of three related social changes: resource management political changes, economic growth, and land tenure reforms. In China, land tenure reforms began with social distribution of land to farming households between 1951 and 1953, followed by a reassignment of all land to people's communes and the establishment of collective forest farms in the early 1960s. Since the early 1980s, exclusive use rights to forests have been granted to households. These included the right to plant trees for private use, but not to convert the land to non-forest use. From 2000, the Guangdong government began to certify forest and land tenure to households and collectives. As a result, since the early 2000s, most forests in Guangdong have been managed by individual households (Chokkalingam *et al.*, 2006).

The Guangdong study does not provide unequivocal evidence about how relevant or important the tenure changes were for forest restoration projects or outcomes. The tenure reforms were heavily constrained, and did not imply a shift to legally recognized private ownership. Farmers gained exclusive right to use the land, but were not allowed to sell the land (Sun *et al.*, 2017). The case suggests that the observed increase in forest cover was the result of multiple national programmes. People joined government-initiated forest restoration efforts because they had positive livelihood outcomes. Specifically, participants in forest restoration were allowed to earn income from selling fruits and other tree products, and could also capture rents when leasing land or engaging in profit-sharing agreements with private enterprises. Based on their evaluation, Chokkalingam *et al.*, (2006) argued that the tenure reforms resulted in rights that could be enjoyed over a long time, and that this was key to forest restoration efforts actually resulting in long-term forest recovery.

The findings of the Guangdong case have been corroborated by more recent studies. For instance, Liu (2015) documents the case of Changting County in western Fujian Province. Tenure reforms not only gave individuals or legal entities wide-reaching land use rights, but also increased the authority of local governments over land and forest matters. Since the 1990s, rights to benefit from tree planting on barren unproductive terrains have improved significantly, and farmers responded by planting tree species that produced marketable fruits. This contributed to tree cover and boosted family incomes. Between 2003 and 2008, almost all of the forestlands that were still held collectively were given under 70-year renewable contracts to individual households, household groups, external investors and village collectives. In practice, this implied that collective management by public entities was replaced by individual management by households (but see a critical discussion by Sun *et al.* (2017) on rights held by individuals and rights still held by township governments). The tenure and decentralization changes coincided with incentives provided to new forestland owners under the Conversion of Cropland to Forest Programme (CCFP), a programme that provided payments for smallholders to plant trees on agricultural land with slopes over 25%. Timber prices increased and forestry taxes dropped correspondingly (Liu, 2015). As a result of these changes, forest cover in Changting County increased from 60% in 1985 to over 80% in 2015. During that period, the average annual farm income increased from US$60 to US$1110.

Vietnam

Vietnam is similar to China in that the country has experienced a forest transition, with a shift from net deforestation to net forest recovery since the mid-1990s. This has resulted in increases in the extent of both natural forests and planted forests (de Jong *et al.*, 2006). Sufficient evidence exists that forest increase was a result of purposeful forest rehabilitation efforts (e.g. de Jong *et al.*, 2006; Dang *et al.*, 2017). Since the late 1980s, Vietnam has implemented land reforms, including forestland reforms. In 1988, the country changed its Land Law, and revised the same law again in 1993 and 1995. The revisions gave households exclusive management rights and responsibility over land and forests therein. A subsequent Law on Protection and Development of Forests further clarified rights and obligations to use, manage and protect forests. These laws were further backed up with government decrees enacted under the country's first major forest landscape restoration programme, Programme 327. The decrees, among others, allocated forests for long-term forestry development to individuals, households and organizations (Dao *et al.*, 2017).

This is the origin of one of the principal mechanisms to grant rights over forestlands, the so-called Forest Land Allocation Policy. Under this policy, Vietnam allocated some nine million ha of land to private users. Many observers suggest that involvement in the programme was low, one reason being the limited rights that were actually granted to local people under the policy (Dang *et al.*, 2017). Following the forest land allocation reforms in 1993, more local forest users have participated in forest restoration (Dang *et al.*, 2017). Forest land allocation has led to a significant increase in forest plantations through smallholder investments, which contributed to the expansion of the country's forest estate and its forest transition (To *et al.*, 2013). However, those who entered into a forest land allocation agreement more often chose an agreement related to production forest plantation, rather than special use or protection forest agreements. A production forest agreement implied financial support by state and private actors, and more liberty to decide the economic goals of forest plantations. The same support and liberties were not available through forest land allocation agreements to rehabilitate special use or protection forests, or for forest land allocation that related to natural production forests (Dang *et al.*, 2017).

Nepal

According to the latest Global Forest Resources Assessment, Nepal has stabilized but not yet increased its forest cover since 2005 (FAO, 2015). In 1992, the country started its Hill Leasehold Forestry and Forage Development (HLFFD) programme, which is also referred to as leasehold forestry. Similarly to other countries in tropical Asia, Nepal has pursued land and forestland reforms, and explored options for synergies between forest restoration and improvement of livelihoods. Lands were nationalized during the late 1950s, and tree cutting was restricted to reduce deforestation, but as a result, forestlands became virtual open access resources. These policies were only reversed in 1978, when the panchayats (administrative bodies of village representatives) were given a dominant role to administer forestlands and forests. This also led to the community forestry projects in the country, in which forests were placed under the control of village councils. The responsibility of village councils was eventually taken over by community forestry user groups (CFUGs). The forest policy changes were formalized in a revised Forest Act of 1993 and Forest Regulations of 1995 (Adhikari *et al.*, 2015). Tenure rights over forests were quite strong, but circumscribed; that is, the rights and obligations of tenure holders were well defined. Tenure rights could only be revoked if an agreed-upon management plan was not adhered to. Community forestry, however, only concerned existing forests; it was not pursued with the primary intention of achieving forest restoration.

In the early 1990s, Nepal saw the actual implementation of leasehold forestry arrangements. Under leasehold forestry, the government transfers land to Leasehold Forestry User Groups (LFUGs). Being poor (considering total annual income and land ownership criteria) is one requirement for membership in a LFUG. Leases are initially given for 40 years and can be extended for an additional 40 years, essentially making LFUGs a form of de facto group private land ownership. LFUGs prepare five-year management plans in consultation with the District Forestry Office. LFUGs receive materials, including seeds, but resource exploitation from leasehold forest lands is restricted, and the management plans need to provide for forest rehabilitation outcomes; for example, LFUG members are obliged to plant and protect trees. The leasehold forestry programme has received development cooperation support and has also been supported by several Nepalese government agencies.

According to Adhikari *et al.* (2015), in 2003, some 1768 LFUGs had been given about 7421 ha of degraded forest lands, but the government was planning to expand the project. The project considerably improved the condition of degraded forests, and it improved the livelihoods of LFUG members (Nagendra, 2007). Apparently, an important outcome that helped with the rehabilitation of forests was the production of fodder, which enabled LFUG members to shift from free grazing to stall feeding. In the locations and period analysed by Adhikari *et al.* (2015), LFUG members increased the number of goats as a result of the project.

Tenure and forest landscape restoration: a complex relationship

The area and condition of forests in forest landscapes in many locations of the world are still declining, as is the condition of forests in those landscapes. Society is increasingly recognizing and valuing the multiple goods and services provided by forests, referred to as 'forest ecosystem services'. Forest landscape restoration needs to play a critical role in assuring that the decline of forests and the reduction in forest ecosystem services provision will be halted. Land tenure and related property rights are key elements of the governance and institutional dimensions of forest landscape restoration. However, the significance of tenure and property rights, and the way they can be harnessed to achieve desired forest landscape restoration outcomes, requires careful consideration.

The various academic and practitioner communities interested in forest development and forest conservation early on recognized the validity of the property rights paradigm (Alchian and Demsetz, 1973), which links the tenure status of an asset with interest in investing in the asset in order to derive benefits. This coincided with the beginning of the debate related to sustainable natural resource use, rural development and nature conservation (Arts *et al.*, 2010). The recognition of the relevance of governance arrangements, including forestland tenure status, to natural resource management by people who depend on those resources for daily subsistence is reflected in the so-called tenure security discourse (Woods, 2010). This discourse has taken hold in the forest landscape restoration literature as multiple tenure and forest landscape restoration narratives.

Forest landscape restoration entails a variety of highly diverse activities in the landscape, such as rehabilitating degraded primary forest, creating biological corridors, establishing forest plantations and promoting agroforestry (Maginnis and Jackson, 2007). Tenure influences these activities differently. Agroforestry, for example, often takes place on land designated as agricultural land. Different tenure regimes will apply to agricultural lands, to land designated for biological corridors, and to lands designated for forestry production. Equally, different rights may be granted to the land, the trees or the products from the trees. It is not uncommon that landscapes where restoration is attempted are characterized by a complex diversity of legal, semi-legal, customary and informal tenure arrangements, which may affect forestlands, nature conservation lands and agricultural lands differently.

Evidence suggests that there is a conceptual relationship between the expected benefits from an investment in land or forest resources and the willingness to invest in longer-term management operations, such as forest restoration or planting trees on agricultural land. Many local actors will only invest in restoration if the benefits as a reward from investment are assured (i.e. revenues from selling timber or NTFPs of trees planted on forest land, or increased water access from restoring forests around water sources). One important question remains in the case of diverse forest landscapes, and that is whether legal tenure can actually guarantee that these benefits will be captured, immediately or in the long run, if investments are made in the property. In our view, the legal tenure–benefit stream linkages are much more complex and nuanced than is reflected in much of the tenure security rights discourse and its multiple contributing narratives. Property rights constitute a conditioning factor between legal tenure and benefits that can be derived from forestlands, or rewards from forestlands that result from investments made. In addition, legal ownership often exists only in name but is not actively enforced by the government, as a result of which legal owners cannot enjoy the rights that legal ownership is supposed to provide. Furthermore, tenure of land, trees and tree products (and services) may differ, exacerbating the complexity of the issue. Rather than emphasizing tenure security, we argue for an increased focus on strengthening the security of property rights over land and resources through recognizing and supporting existing private or communal institutions in a landscape, or otherwise facilitating social capital or good governance.

Several factors restrict the property rights that can be enjoyed by having legal tenure in forest landscapes. Some factors are mainly related to forest land, while others are related to forest and agricultural lands. First, forest land tenure holders generally do not have full freedom to act at their own discretion when deciding on what to do or not to do with or on forest land. The property rights derived from legal tenure are commonly constrained by legislation, regulations and codes that safeguard other actors' needs and demands (Luckert *et al.*, 2011; Sun *et al.*, 2017). Based on the cases discussed earlier, it is also evident that national or subnational governments have the authority to devise and implement policies that constrain forest land holders, or that oblige forest land tenure holders to undertake certain actions. How forest land tenure holders manage forest lands is influenced not only by the rights they actually enjoy but also by policies (Luckert *et al.*, 2011). The rights enjoyed by forest land tenure holders are constrained not only by national or subnational laws but also by policies pursued by national or lower-level governments. The reality in modern societies everywhere is that both tenure holders and non-tenure holders, including forest agencies or the public in general, exercise rights over forest lands. Forest land tenure holders are obliged to share rights and to allow benefits from forestlands to be captured or enjoyed by other actors.

In our assessment, there is as yet insufficient evidence to assume that holding legal forest land tenure is a key incentive or a factor that systematically stimulates investment in forest land restoration. Forest land users will invest in forest restoration if they expect to obtain some benefits from such efforts, and can be sure that when these benefits become available, they, or somebody else whom they favour, will capture those benefits. One of the most important incentives to invest in forest landscape restoration is likely to be economic; that is, a forest land tenure holder will invest in forest restoration if this is likely to yield net economic returns. Other benefits of interest may be social or cultural. However, a decision on whether or not to invest in forest land is quite likely to be influenced by forest land tenure conditions, as is argued by the property rights paradigm. Everything else being equal, actors who consider engaging in forest restoration are likely to do so if and when they are convinced that they, or somebody of their preference, will have the exclusive right to capture the benefits resulted from forest restoration.

Tenure, however, is not the primary driver of forest restoration, although situations can be hypothesized or found in which forest land tenure improvement as an outcome could be the primary incentive (Mansourian, 2016). Forest land tenure, and especially the perceived security of property rights resulting from forest land tenure, will influence the choice of whether to engage in forest restoration or to undertake management efforts that have forest restoration outcomes (e.g. Liu, 2001). Equally, a forest land user, whether or not she or he has forest land tenure, will judge the non-sectoral and policy environment and try to anticipate how that is likely to affect future benefits that are expected from forest restoration efforts or management practices that result in forest restoration.

The four cases of forest landscape restoration discussed earlier, two from China and one each from Vietnam and Nepal, give some idea of the range of possible tenure arrangements and related rights, and how they relate to forest restoration outcomes. Each of the four cases represents successful forest rehabilitation (even if not necessarily achieving all of the criteria one would expect for forest landscape restoration). The tenure arrangements and rights, however, are highly diverse. In none of the four cases was legal tenure granted as freeholder land ownership. Instead, in all four cases, the rights held by the forest users were severely constrained. Importantly, however, in each of the cases, significant economic incentives were provided to make forest restoration attractive to local forest users, while the specific legal tenure status and property rights assured that benefits could be captured into the future.

Conclusions

Legal tenure and related property rights are relevant and important in forest landscape restoration, and are prerequisites for the effectiveness of other incentives for promoting forest restoration. However, the assumption that assuring legal tenure with full property rights is a decisive incentive in and of itself for forest restoration is not supported by empirical evidence. Legal tenure status over forest lands and property rights derived from legal tenure are diverse between countries, and even within countries. It is possibly of little relevance to define specific principles, let alone blueprints of what exactly could be the best legal tenure or rights to support or facilitate forest landscape restoration. It is evident in each of the empirical cases we reviewed that legal tenure and property rights were highly constrained for forest users, while forest restoration was successfully implemented.

The case studies show that even when legal tenure is held over forest land, the property rights derived from tenure may be severely constrained and not necessarily widely recognized, respected or protected. Legal tenure can constitute a less desired option in terms of rights that are granted, and other arrangements, that is, customary arrangements or other forms or non-statutory enforced arrangements, may result in more secure rights. Granting some kind of legal tenure is often a part of successful forest landscape restoration, but arrangements leading to successful outcomes are commonly variable. When forest landscape restoration is desired, legal tenure needs to be addressed, but what kind of tenure is provided to forest users, what rights the tenure implies, and what constraints are the most appropriate all depend very much on the wider legal, political and social context in which forest restoration is being pursued.

Note

1 www.iucn.org/theme/forests/our-work/forest-landscape-restoration.

References

Adhikari, B.R., Baral, N.R., Hancock, J., Kafley, G., Koirala, P., Reijmerinck, J. and Shapiro, B. (2015) *Regenerating forests and livelihoods in Nepal: A new lease on life*. FAO and IFAD, CABI, Wallingford, UK.

Alchian, A.A. and Demsetz, H. (1973) 'The property right paradigm', *The Journal of Economic History*, vol 33, no 1, pp. 16–27.

Appanah, S., Shono, K. and Durst, P.B. (2015) 'Restoration of forests and degraded lands in Southeast Asia', *Unasylva*, vol 16, no 3, pp. 52–63.

Arts, B.J.M., Appelstrand, M., Kleinschmit, D., Pülzl, H., Visseren-Hamakers, I.J., Eba'a Atyi, R., Enters, T., McGinley, K. and Yasmi, Y. (2010) 'Discourses, actors and instruments in international forest governance'. In Rayner, J., Buck, A. and Katila, P. (eds) *Embracing complexity: Meeting the challenges of international forest governance. A global assessment report*. International Union of Forest Research Organizations, Vienna, Austria, pp. 57–74.

Berrahmouni, N., Parfondry, M., Regato, P. and Sarre, A. (2015) 'Restoration of degraded forests and landscapes in drylands: guidelines and way forward', *Unasylva*, vol 16, no 3, pp. 37–43.

Besley, T. (1995) 'Property rights and investment incentives: theory and evidence from Ghana', *The Journal of Political Economy*, vol 103, no 5, pp. 903–937.

Biermayr-Jenzano, P., Kassam, S.N. and Aw-Hassan, A. (2014) *Understanding gender and poverty dimensions of high value agricultural commodity chains in the Souss-Masaa-Draa Region of South-Western Morocco. Working paper, mimeo*, ICARDA, Amman.

Bruce, J.W. (1998). *Review of tenure terminology, Tenure Brief*, No. 1, University of Wisconsin, Madison.

Cano, W., de Jong, W., Zuidema, P. and Boot, R. (2014) 'Diverse local regulatory responses to a new forestry regime in forest communities in the Bolivian Amazon', *Land Use Policy*, vol 39, pp. 224–232.

Chang, H.J. (2011) 'Institutions and economic development: theory, policy and history', *Journal of Institutional Economics*, vol 7, no 4, pp. 473–498.

Chokkalingam, U., Sabogal, C., Almeida, E., Carandang, A.P., Gumartini, T., de Jong, W., Brienza, Jr., S., Meza Lopez, A., Murniati, Nawir, A.A., Rumboko, L., Toma, T., Wollenberg, E. and Zhou, Z. (2005) 'Local participation, livelihood needs, and institutional arrangements: three keys to sustainable rehabilitation of degraded tropical forest lands'. In Mansourian, S., Vallauri, D. and Dudley, N. (eds) *Forest restoration in landscapes: Beyond planting trees*, Springer, New York, pp. 405–414.

Chokkalingam, U., Zhou, Z., Wang, C. and Toma, T. (2006) *Learning lessons from China's forest rehabilitation efforts: National level review and special focus on Guangdon province*. CIFOR, Bogor, Indonesia.

Cotula, L. and Mayers, J. (2009) *Tenure in REDD – Start-point or afterthought? Natural Resource Issues No. 15*, International Institute for Environment and Development, London.

Cousins, B. (2005) 'Tenure reform in South Africa: titling versus social embeddedness', *Forum for Development Studies*, vol 32, no 2, pp. 415–442.

Cronkleton, P. and Larson, A. (2015) 'Formalization and collective appropriation of space on forest frontiers: comparing communal and individual property systems in the Peruvian and Ecuadorian Amazon', *Society and Natural Resources*, vol 28, pp. 496–512.

Dang, T.K.P., Visseren-Hamakers, I.J. and Arts, B. (2017) 'The institutional capacity for forest devolution: the case of forest land allocation in Vietnam', *Development Policy Review*, pp. 1–22.

Dao, M.T., Yanagisawa, M. and Kono, Y. (2017) 'Forest transition in Vietnam: a case study of Northern mountain region', *Forest Policy and Economics*, vol 76, pp. 72–80.

de Jong, W. (2018) *Progress and Challenges in Meeting the Forest Landscape Restoration Commitments in Asia*, Asia Pacific Network for Sustainable Forest Management and Rehabilitation, Beijing, China.

de Jong, W., Sam, D.D. and Hung, T.V. (2006) *Forest Rehabilitation in Vietnam. Histories, Realities and Futures*, CIFOR, Bogor, Indonesia.

de Jong, W., Liu, J. and Youn, Y-C. (2017a). 'Land and forests in the Anthropocene: trends and outlooks in Asia', *Forest Policy and Economics*, vol 79, pp. 17–25.

de Jong, W., Galloway, G., Katila, P. and Pacheco, P. (2017b) 'Forestry discourses and forest based development – an introduction to the Special Issue', *International Forestry Review*, vol 19 (S1), pp. 1–9.

FAO (2015) *Global Forest Resources Assessment, 2015*, Food and Agricultural Organization, Rome.

Feder, G. and Feeny, D. (1991) 'Land tenure and property rights: theory and implications for development policy', *The World Bank Economic Review*, vol 5, no 1, pp. 135–153.

Fortmann, L. and Bruce, J. (1991) *You've got to know who controls the land and trees people use: Gender, Tenure and the environment*, IDS, Sussex.

Galik, C.S. and Jagger, P. (2015) 'Bundles, duties, and rights: a revised framework for analysis of natural resource property rights regimes', *Land Economics*, vol 91, no 1, pp. 76–90.

Haley, D. and Luckert, M.K. (1992) 'Property rights and the management of forest resources: the Canadian experience', *Commonwealth Forestry Review*, vol 71, no 2, pp. 91–94.

Katila, P., Galloway, G., de Jong, W., Mery, G. and Pacheco, P. (2014) *Forests under pressure: Local responses to global issues*, IUFRO, Helsinki.

Kerekes, C.B. and Williamson, C.R. (2010) 'Propertyless in Peru, even with a government land title', *American Journal of Economics and Sociology*, vol 69, no 3, pp. 1011–1033.

Laestadius, L., Buckingham, K., Maginnis, S. and Saint-Laurent, C. (2015) 'Before Bonn and beyond: the history and future of forest landscape restoration', *Unasylva*, vol 16, no 3, pp. 11–19.

Lamb, D. (1994) 'Reforestation of degraded tropical forest lands in the Asia-Pacific region', *Journal of Tropical Forest Science*, vol 7, no 1, pp. 1–7.

Lamb, D., Erskine, P.D. and Parrotta, J.A. (2005) 'Restoration of degraded tropical forest landscapes', *Science*, vol 310, p. 1628.

Lawry, S. (2015) *How does land tenure affect agricultural productivity? A systematic review*, http://blog.cifor.org/26908/land-tenure-reforms-africa-review?fnl=en.

Leipold, S. (2014) 'Creating forests with words – a review of forest-related discourse studies', *Forest Policy and Economics*, vol 40, pp. 12–20.

Liu, D. (2001) 'Tenure and management of non-state forests in China since 1950: a historical review', *Environmental History*, vol 6, no 2, pp. 239–263.

Liu, J. (2015) *From ecological construction to ecological civilization – The case of changting from the perspective of social sciences*, China Social Sciences Publishing House, Beijing.

Luckert, M.K. (1991) 'The perceived security of institutional investment environments of some British Columbia forest tenures', *Canadian Journal of Forest Research*, vol 21, no 3, pp. 318–325.

Luckert, M.K., Haley, D. and Hoberg, G. (2011) *Policies for sustainably managing canada's forests: Tenure, stumpage fees, and forest practices*, UCB Press, Vancouver.

Maginnis, S. and Jackson, W. (2007) 'What is FLR and how does it differ from current approaches?' In Rietbergen-McCracken, J., Maginni, S. and Sarre, A. (eds) *The forest landscape restoration handbook*, Earthscan, London, pp. 5–20.

Mansourian, S. (2016) 'Understanding the relationship between governance and forest landscape restoration', *Conservation and Society*, vol 14, pp. 267–278.

Mansourian, S. (2017) 'Governance and forest landscape restoration: a framework to support decision-making', *Journal for Nature Conservation*, vol 37, pp. 21–30.

Mansourian, S., Vallauri, D. and Dudley, N., eds. (2005) *Forest restoration in landscapes: Beyond planting trees*, Springer, New York.

Mansourian, S., Razafimahatratra, A., Ranjatson, P. and Rambeloarisao, G. (2016) 'Novel governance for forest landscape restoration in Fandriana-Marolambo, Madagascar', *World Development Perspectives*, vol 3, pp. 28–31.

Nagendra, H. (2007) 'Drivers of reforestation in human-dominated forests', *Proceedings of the National Academy of Sciences* vol 104, no 39, pp. 15218–15223.

Payne, G. (2004) 'Land tenure and property rights: an introduction', *Habitat International*, vol 28, no 2, pp. 167–179.

Sabogal, C., Besacier, C. and McGuire, D. (2015) 'Forest and landscape restoration: concepts, approaches and challenges for implementation', *Unasylva*, vol 16, no 3, pp. 3–10.

Schlager, E. and Ostrom, E. (1992) 'Source property-rights regimes and natural resources: a conceptual analysis', *Land Economics*, vol 68, no 3, pp. 249–262.

Sun, V., Lin, J. and Chand, R.C.K. (2017) 'Pseudo use value and output legitimacy of local growth coalitions in China: a case study of the Liede redevelopment project in Guangzhou', *Cities*, vol 61, pp. 9–16.

Sunderlin, W.D., Larson, A.M. and Cronkleton, P. (2009) 'Forest tenure rights and REDD'. In Angelsen, A. (ed.) *Realising REDD*, CIFOR, Bogor.

To, X.P., Tran, H.N. and Zagt, R. (2013) *Forest land allocation in Viet Nam: Implementation processes and results*, Info Brief Tropenbos International, Hanoi, Vietnam.

Von Benda-Beckmann, F., von Benda-Beckmann, K. and Wiber, M. (2006). 'The properties of property'. In von Benda-Beckmann, F., von Benda-Beckmann, K. and Wiber, M. (eds) *Changing properties of property*, Berghahn, New York, pp. 1–39.

Wilson, S.J. and Cagalanan, D. (2016) 'Governing restoration: strategies, adaptations and innovations for tomorrow's forest landscapes', *World Development Perspectives*, vol 4, pp. 11–15.

Woods, S. (2010) 'Tenure security discourse. An analysis of international and South African perspectives on formalising land tenure'. University of Cape Town, http://landlawwatch.co.za/download/2010/CR/Samantha%20Woods-CR-Tenure%20Security%20discourse%20.pdf.

11 Polycentric governance and forest landscape restoration

Considering local needs, knowledge types and democratic principles

R. Patrick Bixler, Theresa Jedd and Carina Wyborn

Introduction

Forest landscape restoration (FLR) holds a unique position in the realm of environmental stewardship and can serve as a model for collective environmental action. A host of expertise is required to bring the best available scientific, traditional and experiential knowledge to bear on project implementation. Restoration requires reconciling differing and sometimes conflicting values regarding forest resources and gaining agreement on the future desired condition of the restoration activity, as well as integrating different knowledge bases and scientific perspectives that guide restoration activities. In many cases, the necessary revenue to cover the cost of restoration is not generated from forest by-products, requiring an investment of capital from an external source: either a governmental entity, an international institution or an innovative market mechanism. Finally, forest landscapes provide a range of services – timber resources, fuel stocks, food sources, water filtration, erosion control, aesthetic viewsheds, recreational opportunities, cultural and spiritual benefits, wildlife habitat, carbon capture and storage – that are prioritized differently depending on the stakeholder perspective. This creates an array of actors and authorities that have a 'stake' in the restoration process (Mansourian, 2017).

These factors shape the contours of restoration decision-making and determine the feasibility of FLR. Governance refers essentially to decision-making, which was traditionally the remit of governments. However, the shift from government-based management to polycentric governance highlights the contemporary multi-actor and multi-purpose natural resource reality (Lemos and Agrawal, 2006; Newell *et al.*, 2012). In some natural resource management settings, a single actor, typically a state or federal bureaucratic agency, is designated to restore resources in accordance with the prevailing biocentric perspective favoured by the agency. In these situations, the lines of authority and the administrative boundaries within which decision-making occurs are often clearly defined. On the other hand, multi-stakeholder governance is less formalized, more difficult to control, and involves a diverse set of actors. This broader context of environmental governance encompasses the set of regulatory processes, mechanisms and organizations through which political actors influence actions and outcomes (Lemos and Agrawal, 2006). Adaptive governance is an umbrella phrase for collaborative and participatory alternatives to top-down decision-making (Dietz *et al.*, 2003) that emphasizes several key dimensions related to governance, knowledge and action (Wyborn, 2015b). Adaptive governance is characterized by continual generation and integration of knowledge, social learning and refinement of approaches based on new information; flexible institutions and multi-level governance to foster shared responsibility and collaboration within a social network; and development of adaptive capacity to address uncertainty and change (Folke *et al.*, 2005; Jacobson and Robertson, 2012). This lineage of governance theory informs perspectives on polycentric governance.

Polycentric and multi-level governance are relatively newer ways to understand forest governance, including restoration activities. These systems-level ways of thinking seek to characterize the multiplicity of actors and modes of governance operating in diverse and overlapping spheres of authority (Bixler, 2014). This perspective starts from the premise that no fixed spatial or temporal level is appropriate for governing ecosystems and their services sustainably, effectively and equitably. Thus, it argues for conceptualizing and designing collective action systems that are adaptable, take into account multiple perspectives, and foster horizontal and vertical linkages between agencies, organizations and institutions.

Understanding the nature of vertical linkages between decision-making units and multiple spheres of authority is important. This is particularly the case when adapting forest restoration policies and projects from large development institutions such as the World Bank, or country-level agencies such as the Deutsche Gesellschaft für Internationale Zusammenarbeit (the German Corporation for International Cooperation) or the Agence Francaise de Developpement (the French Development Agency) to local needs. In such cases, these large agencies seek public input in order to improve the uptake of new practices and technologies. In other cases, local community needs are directly expressed in an upward fashion: when communities or groups of land owners convene to discuss best management practices, for example, they can affect policy decisions at a larger landscape scale. The literature on forest governance has focused on knowing whether and how these dynamics are successful. The effectiveness of forest restoration can be evaluated using indicators such as policies, plans, monitoring activities, practices and law enforcement (World Resources Institute, 2013) and considering dimensions of effectiveness, efficiency and equality of the activities (Bixler *et al.*, 2016). Similarly, the adaptation literature highlights a need for consideration of community needs when engaging in projects aimed at climate adaptation; for example, when considering development projects, forestry should be placed within a host of other land uses, such as farming, fishing and livestock grazing (Thomas and Twyman, 2005). When considering these complementary and overlapping livelihoods in practice, it becomes clear that scholars and practitioners need a consistent analytical framework.

To be successful, the governance dimensions of FLR must be front and centre. As Mansourian notes, 'who decides what and where to restore? How are all stakeholders engaged? Who benefits? Who loses? How are benefits transferred? These are key challenges in forest landscape restoration projects' (Mansourian, 2016). Problems can ensue when there is a mismatch in resources or social group needs are not met. Conflicts can arise when there is a misalignment of resource management boundaries, ecological boundaries and administrative boundaries. Communication, coordination and collaboration, if aligned through the appropriate governance framework, can address the scale mismatch challenge and work to integrate science and values on FLR projects. For example, in many locations, communities use wood biomass for fuel to cook meals and warm their homes. Simply advocating for an increase in forest cover without adequate programmes to provide forest-dwelling communities with alternative fuel sources may fail, as it does not consider local needs. When restoring dry woodland ecosystems in the Punjab province of Pakistan, Gratzfeld and Khan (2015) found that forest authorities were often pitted against competing interests of local communities, and in particular, landless populations. One tactic to reconcile these needs was to recognize the range of forest product uses. In the Punjab region, bringing dry forests to the attention of the scientific and conservation communities involved calling attention to the use of scrub species, including acacias and evergreen species, that are used for crucial fuel products and also less recognized activities like toothpaste production and basketmaking (Gratzfeld and Khan, 2015).

In this chapter, we will describe polycentric governance and draw on empirical evidence to illustrate how it is a practical framework for the governance of FLR. Through a short review of polycentric governance literature, we will discuss and define the concept. We will highlight key tenets that distinguish polycentric governance from other forms: the existence of many centres for decision-making that operate on multiple levels; a common system of rules (be they institutionally or culturally enforced); and the incorporation of different values and beliefs (including scientific perspectives) within a system that emphasizes learning and feedback. We illustrate these attributes of polycentric governance by drawing on forest restoration case studies generated from our own research and experiences as well as examples from the published literature. With appropriate consideration of governance scales, large-scale projects can be long lasting. If technical solutions (which may seem appealing as a rapid fix) are applied in a top-down fashion, they can mask underlying concerns that may crop up to undermine restoration in the mid- to long term. Although there are many ways to achieve collective action on forest restoration, polycentricity is considered unique because it is capable of accounting for decisions made across a broad range of actors and across multiple levels of organization while integrating varied values and knowledge bases. Rather than suggesting that polycentricity is a panacea, we argue that it is a helpful frame to advance the design and implementation of forest restoration around the world.

What is polycentricity?

In the early 1960s, the first polycentric studies focused on collective goods in metropolitan areas. These studies empirically observed that overlapping jurisdictions with multiple loosely linked centres of authority provided better services than one centralized unit or many small units with distinct territories (Ostrom *et al.*, 1961). This work led to the conceptualization of polycentric governance systems that have multiple centres of authority, or multiple governing authorities at differing scales rather than a monocentric unit (Ostrom, 1999). Such systems create opportunities for local institutions to evolve by tightening monitoring and feedback loops and enhancing associated institutional incentives in ways that improve the fit between knowledge, action and social-ecological contexts (Berkes *et al.*, 2003). This is particularly salient in restoration efforts, as it allows those closest to the resource, who have a sense of local needs and strategies, to play a role, while at the same time tempering this embedded perspective with knowledge and resources from elsewhere. In this context, polycentric networks can facilitate and connect multiple centres of authority across different institutional levels, providing communication channels for the various stakeholders at multiple levels and enabling various hubs or centres of decision-making (Berkes, 2010; Bodin and Prell, 2011). Across the different domains (or issue areas related to forest restoration), the various groups can pitch in when their expertise or resources are needed (Armitage *et al.*, 2007). In the framework of FLR, this relieves pressure on any one actor or organization to take on the task of restoring a degraded forest landscape on their own, and facilitates governance arrangements that enable different groups or communities to have a voice in shaping restoration decision-making processes.

Each unit within a polycentric system exercises considerable independence in the making of norms and rules within a specific decision-making unit (such as a family, firm, local government, network of local governments, state or province, region, national government or international regime), yet the polycentric system is guided by a larger set of institutional or cultural rules of interaction. This differentiates a coordinated system of governance from individual actions. Forest degradation occurs in all parts of the world and various governing contexts, yet severe forest degradation often occurs in states governed by illiberal regimes. INTERPOL (2016) found a general correlative trend between corruption and annual rates of deforestation; in this study, 9 out of the 12 country cases that struggled to control corruption also experienced net deforestation. The same study concluded that forestry projects are particularly susceptible to bribery and fraud perpetrated by officials in forestry agencies. Through a polycentric lens, keeping various actors engaged in forest processes builds in needed oversight and social sanctions, and builds the social legitimacy and acceptance required for local uptake. Figure 11.1 highlights these components.

As coordinated frameworks for action, polycentric systems have considerable advantages, given their mechanisms for mutual monitoring, learning and adaptation strategies over time (Ostrom, 2010). As Ostrom (1999) highlighted, although polycentric governance systems are frequently criticized for being too complex, redundant and lacking central direction when viewed from the static, simple-systems perspective, they have considerable strength when viewed from the dynamic, complex-systems perspective, particularly when considering a system that is concerned with the vulnerability of governance systems to disturbances.

Polycentricity: multiple centres of decision-making across levels and scales

We advocate for a polycentric approach because forest restoration in landscapes is about more than the forestry practices and techniques used to carry out projects. Restoration should encompass broader social goals and be politically sensitive to the cultures and jurisdictions within which it takes place. Scholars of FLR call for a sincere appreciation of temporal and spatial scales while building in a range of stakeholder perspectives (Mansourian, 2017). Polycentric systems are defined by the characteristic that they exhibit decision-making capability at different levels (e.g. local, state, federal, international).

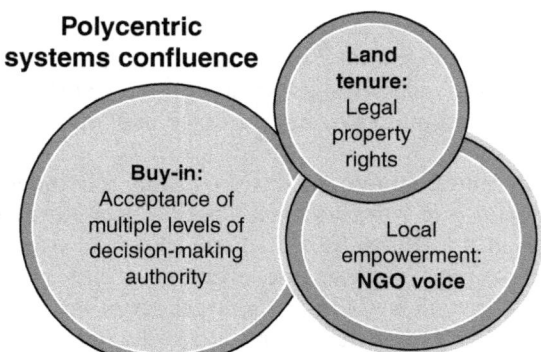

Figure 11.1 The governance requirements of polycentric forest restoration involve local land rights and empowerment, accompanied by an acceptance of authority across levels of governing.

Building a socially driven stakeholder needs assessment into the beginning of a project is a critical element to ensure success. Orchard and Stringer (2016) conducted a large-scale project in Swaziland that involved an agroforestry component which established new seedling nurseries and required current grazing areas to be fenced to prevent damage to the plots. In Swaziland, a stratified socio-economic system coupled with an absolute monarchy made previous development efforts difficult (Orchard and Stringer, 2016). Without a polycentric approach, the managers in this project-as-study felt that it would have been impossible to understand the complex local institutional rules while acknowledging the project funders' needs (Orchard and Stringer, 2016). To navigate the various policy settings and jurisdictional requirements, it quickly becomes apparent that a nuanced understanding of a governance context is required. Polycentricity can provide a framework for making sense of these complex and overlapping policy architectures.

Local-level actors may pick up more quickly on environmental signals indicating the need for restoration and will most likely be called upon for restoration implementation. Yet, where restoration needs and plans cross boundaries, local-level actors often lack the capacity and authority, and thus, higher-level decision-makers in the polycentric system need to be involved. Similarly, many local-level decisions for restoration may be driven by national or international target setting. In the case of higher-level decisions driving local-level responses, regularly engaging local agencies and citizen groups is critical.

Bottom-up engagement is not only more ethically sound but can result in better environmental outcomes. In various settings around the United States, DellaSala *et al.* (2003) found that citizen activism can be a critical component in facilitating the uptake of ecologically sound practices during forest restoration, such as road recontouring to restore hydrologic function and tree thinning to reduce wildfire fuel loads. In many cases, citizen groups provide innovative ideas for solving sticky problems and the on-the-ground workforce to carry out projects. Unfortunately, some barriers prevent project leads from involving local participants. Misperceptions about local capacity can hinder internationally led efforts from truly engaging local participants.

In Moroccan semi-arid forest restoration projects, Derak *et al.* (2017) argue that a participatory approach has been uncommon because of the challenges that poverty and illiteracy pose to long-term project maintenance. However, a recent success case in Northern Morocco's Béni Boufrah valley illustrates the benefits of citizen involvement. In this region, where the Barbary red cedar (*Tetraclinis articulata*) has declined due to human settlement pressures, residents had an aversion to *Pinus* plantings, which were often used in reforestation efforts (Derak *et al.*, 2017). Using locally important species (e.g. Barbary red cedar and mastic (*Pistacia lentiscus*), valued for its resin), a resident-driven project had 60% and 90% survival rates, respectively, attributable to the lack of vandalism and grazing two years after the initial planting (Derak *et al.*, 2017). This example speaks to the importance of citizen involvement, especially in settings where governance dynamics may not routinely stand up to social and economic pressures.

Governance arrangements in polycentric systems are more effective when citizens have the ability and are authorized to self-organize not only one but multiple governing authorities of differing scales (Ostrom *et al.*, 1961; Ostrom, 1991, 1997). For example, in North America, noxious weeds and exotic fish species have invaded terrestrial and aquatic ecosystems, and thousands of miles of abandoned logging roads intersect key wildlife habitat. This landscape fragmentation can lead to increased sediment in rivers that provide drinking water to population centres and in blue ribbon trout streams, impacting recreation. Erosion from mining necessitates focused and expensive cleanup efforts. Decades of fire suppression – a management response to catastrophic fires in the early twentieth century – have dramatically altered the ecology of western forested ecosystems and resulted in unnaturally high accumulation of forest fuels (Westerling, 2016). Addressing the numerous restoration challenges in these landscapes explicitly challenges the notion of scale. Ecological processes operate across spatial scales, while climatological patterns govern localized processes (e.g. seedling recruitment). All the while, small-scale patterns (stand level structure of forest patches) scale up to the collective patterns of fire behaviour and the resulting emergent properties of landscape composition and arrangement. Working across scales requires a forest restoration response focused on the landscape level, engaging a broad range of actors with their knowledge and experience.

In a polycentric governance system, each unit exercises considerable independence to make and enforce rules within a circumscribed scope of authority for a specified geographical area. In such a system, some units are general purpose, while others may be highly specialized. Self-organized resource governance systems may be special districts, private associations, or parts of a local government. These are nested in several levels of general-purpose governments that also provide conflict resolution mechanisms (Andersson and Ostrom, 2006). Polycentric approaches to governance are advantageous because they build in the flexibility that is necessary to carry out work at various scales and to address major challenges to project implementation. In the case of forest restoration, they also make sense. It is difficult to imagine trying to achieve a reforestation project without community involvement to help plant the seedlings, for example.

Forest restoration activities require a broad base of skills to carry out, and their approval is based upon perceptions of how necessary or effective they have been. A study of Finnish forestry practices found that clear cutting and thinning activities were viewed less favourably from an aesthetic perspective, while undergrowth removal had a positive recreational value, but a natural state was valued most for aesthetics (Tahvanainan *et al.*, 2001).

In the United States, typical forest restoration practices occur at the stand level: from tree planting, to stream channel repairs, to road decommissioning, to prescribed burns. While it may seem that, because these projects take place at the community level, involvement from other levels of government is irrelevant, that could not be further from the truth. Consider the practice of stream channel alteration. For some cold-water fish species, protective management measures will include the creation of deep-water pools to maintain the needed pockets of cooler temperatures. However, local managers rely on information produced by scientific research to inform their decisions as to which tools to use and where to place the pools. Polycentric arrangements can help connect needed science-based guidance with the local and experiential knowledge of local implementation staff. A polycentric approach connects local knowledge back up to large research operations and governance decisions being taken at broader scales. Figure 11.2 demonstrates the interconnected nature of local

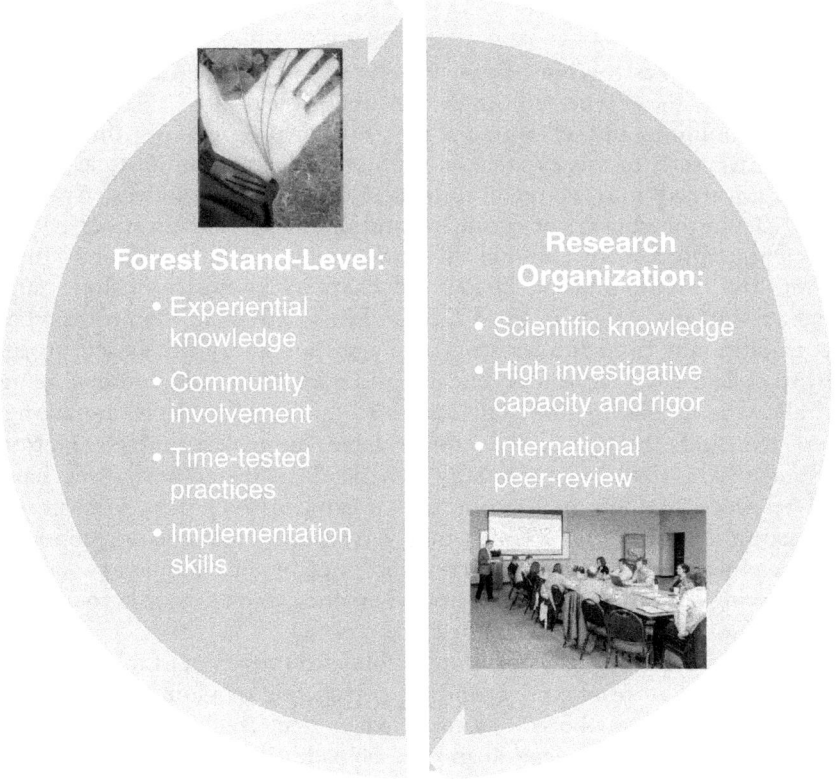

Figure 11.2 Interface of knowledge and scale: polycentricity pairs forest stand-level knowledge with science from research organizations.

Source: photo © Eric North and Public Policy Center, University of Nebraska – Lincoln.

activities and research products in an iterative fashion. As science products are produced, they are transferred back into local contexts, which are also sites of expertise themselves.

Bringing democratic principles into restoration: key considerations

The process of bringing in a mix of actors needed for forest restoration can fundamentally reshape the governance landscape, creating equitable and fair arrangements that consider a range of social needs. Polycentric approaches enable different actors to have 'voice', nesting them within decision-making processes at relevant scales. The ability to contribute to ongoing decision-making support for implementation keeps participants engaged.

Accountability

Polycentricity blurs the lines between who is making the decisions about the landscape, those who will implement the projects, and those who live in the communities and will monitor the project's progress over time. Political scientists refer to this expansion of decision-making authority as bringing 'the governed' into the realm of being 'governors' themselves (Avant *et al.*, 2010). In the context of forest restoration, this opens up space within the scientific process for community involvement. As other authors (Humphreys, 1999; Gulbrandsen, 2012) and chapters in this book have suggested, a shift took place in the 1990s and early 2000s in the process that was traditionally used (under strict management and yields-based protocols) to collect data about forest health and monitor trends. In opening up and allowing a role for non-state actors to take part in decision-making, forest managers have received added value in project implementation assistance. This has been mutually beneficial, as local businesses have received access to additional resources that can help keep them in business. For example, timber mills, which may have been experiencing economic downturn because production was not sustained by local supply, benefitted from serving as restoration partners when they gained access to timber harvest products (Schultz *et al.*, 2012).

This process has brought non-state actors into the realm of making decisions about public resources. From the perspective of maintaining democratic integrity, it can be troublesome when unelected individuals make decisions that affect the general public. Outside of the ethical challenges that citizen participation poses, it can introduce an undue burden for administrators (Cupps, 1977). Bringing citizens into the scientific process has been argued to introduce scientific uncertainty about natural resource management (Gerlak and Heikkila, 2006). We argue that using citizen and stakeholder input in the design phase of forest restoration may lead to improved outcomes, because it gives community members a chance to design the hypotheses that are tested and ultimately to inform the techniques used in project implementation.

Polycentricity brings in a mix of different actors, but also fundamentally changes the way they work together. An informal, flexible and open-ended setting replaces codified rules and prescribed procedures. Some have pointed to a trade-off between flexibility and accountability – suggesting that as processes are more flexible (subject to fewer obligations), they struggle with accountability in the eyes of the public. When collaborative arrangements do not correct for power imbalances but reproduce pre-existing power dynamics, it can be difficult for members of the general public to determine how they can make their voices heard in a large spatial setting (Lau, 2014). Some have critiqued polycentric governance as lacking clear mechanisms for accountability, because these networked forms of governance can allow 'gatekeepers' to restrict access to information and resources (Lau, 2014). On the other hand, traditional, hierarchical governance has top-down flows of authority. When the governance network encounters conflict, it is easier to make a clear assignment of blame and to directly sanction violators. For example, it is possible to 'vote out' those in elected positions, or pursue legal grievances through judicial avenues. However, under an arrangement where authority is shared, it's not always clear how power is checked.

At the local level, courts are often used to enforce legal institutions. This process allows affected individuals and groups to register their complaints and be compensated for their damages. Judicial bodies, though, at the international level are plagued with problems related to opt-out and the barrier of state sovereignty. When forest restoration takes place in a transboundary landscape, it may be necessary to consider how to incorporate the interests of the key agencies and organizations outside the court system. When competing interests are represented in a forested ecosystem, how can those whose voices are underrepresented be heard? Polycentric arrangements should incorporate a consideration of channels for holding decision-makers accountable to their stakeholders. Depending on the degree of flexibility, it is possible that non-governmental organizations (NGOs) may be more or less accountable. For example, when private resources are put towards projects, implementation may depend less upon the direct efforts of public officials, placing more impetus on civil society to monitor outcomes (Newman, 2004; Lau, 2014) or embed accountability into network processes (Jedd and Bixler, 2015).

Polycentric governance shifts the location of authority and control away from the state. With its decentralized nature, it challenges conventional notions of power in international relations (Kahler, 2009), bringing new actors to the table. In the case of forested ecosystems in North America, this has historically been a beneficial process, allowing larger groups to draw upon the experience of smaller, more local groups (Pedynowski,

2003). Membership in roundtable processes is voluntary, and civil society groups are brought into the decision-making process. The nonbinding nature of commitments is a partial necessity, since agencies do not formally give up authority. In the case of the Roundtable on the Crown of the Continent (see Wyborn and Bixler, 2013; Jedd and Bixler, 2015; Bixler *et al.*, 2016, 2018), accountability is enforced by each individual agency and organization through informal mechanisms (Jedd and Bixler, 2013). For example, in the United States, the National Park Service can participate in a polycentric governance network, but at the end of the day, it is very clear about having its own agency mission and mandate. The techniques used for conservation within Park boundaries must be cleared through federal policies and procedures. Thus, we suggest that in high-capacity states with limited corruption, accountability can be retained in a polycentric arrangement by referring back to the landowning entity within each jurisdiction.

In other institutional contexts, mechanisms for accountability in the context of polycentric governance are evolving. Bringing in local community partners may amplify the longer-term success of restoration projects. Here, NGO and citizen groups can be key partners in building accountability systems. For example, in North Africa, reforestation is being used as a tool to halt the expansion of the Sahara desert. However, these efforts are being challenged for both governance and ecological reasons (see Schilling *et al.*, 2012; Bentley *et al.*, 2015). In the Moroccan dry forest example (see earlier), the dryland species declined over time due to desertification trends and increasingly arid conditions. Perhaps more of a challenge, the smaller scrub species that can be used for a range of purposes were not valued as economic commodities, and the smaller *Pinus* species that were used in replanting efforts to replace the red cedar (Derak *et al.*, 2017) were found to interfere with subsistence agriculture and livestock grazing. The vandalism and corruption challenges to the red cedar restoration, which on the surface were driven by governance failures, reflect local aversion to particular species. Identifying the drivers of deforestation can go far in ameliorating local concerns (Hosonuma *et al.*, 2012). Additionally, when communities are brought into the process of selecting and planting tree species, the long-term outcomes may be greatly improved. The added benefit is that trust may also be increased within the citizenry.

Legitimacy

Legitimacy in the context of forest restoration refers to the process through which policy becomes accepted. Cadman (2011) writes that legitimacy is the standard through which governance outputs should be judged. In forest governance, certification programmes have contrasting legitimacy mechanisms, and vary depending on whether their main contributions come from civil society actors (e.g. the Forest Stewardship Council) or whether they are initiated by market-based actors driven by economic interests (e.g. the

Programme for the Endorsement of Forest) (Cadman, 2011). In these cases, the basis of acceptance is dependent on the original stakeholders' interests. Acceptance and buy-in for forest practices is a process that takes place in repeated interactions. Bringing stakeholders together for a restoration project in a committed series of engagements can reduce conflict at a later date. Nie and Metcalf (2015) suggest that the historically adversarial nature of interactions between citizen groups and US federal agencies (namely, the US Department of Agriculture (USDA) Forest Service) formed the bedrock of American environmental politics at a time when citizen voices were needed as a check on the power of governmental agencies to determine outcomes on forested lands. When collaboration is written into federal legislation (Schultz *et al.*, 2012), stating that restoration objectives should be achieved with NGO input and implemented with private partners, the role of litigation and conflict may be diminished. Indeed, it would be difficult to imagine a lawsuit emanating from a group that is directly involved in writing a forest plan.

In the US context, legitimacy has shifted. A sustained yield perspective formed the basis of legitimacy for government-driven, top-down forest management. Under sustained yield, decisions about forests were made to encourage timber production above non-material, aesthetic values. Legitimacy rested with the state, which was seen as the primary avenue for pursuing the public good (Scott, 1998). Land and forest managers, often urged by federal funding programme structural shifts, are adopting a 'new' legitimacy, or one that is based upon shared authority in (bottom-up) collaborative arrangements (Schultz *et al.*, 2012).

This basic observation represents the need to develop methods for incorporating community needs from the beginning into policy processes, and has also been a fundamental explanatory factor for success in community forest initiatives in the developing country context. In Nepal, for example, the move towards decentralization was coupled with forest restoration, serving as an example of how dynamic transformations of governance can modify or reform forestry practices. As governance is drawn closer to the level of a forest stand, local voices are heard and community members feel empowered to make decisions about restoration. There is wide variation among decentralization processes throughout the world, with varying levels of control granted to forest-dwelling communities. Agrawal and Ostrom (2001) argue that local empowerment through NGO networks is important, along with land rights reform. The community forestry movement in Nepal is often cited as an example of the positive outcomes of this nexus of empowerment and land rights transformation.

This shift towards including non-state actors in forest governance has resulted in net gains in capacity and local buy-in. When development networks are adept at navigating local customs and practices, their inclusion broadens the base of legitimate participants and explains why polycentric

Box 11.1 Community empowerment through decentralization in Nepalese forestry

The Kingdom of Nepal extends 800 km east to west along the southern slopes of the Himalayas in South Asia. More than 80% of the area is covered by rugged hills and mountains, including Sagarmatha (Mt. Everest) and seven other of the world's highest peaks. These areas harbour biological riches of both the Indo-Malayan and Paleoarctic zones, including many endemic Himalayan floral and faunal species. About 32% of Nepal's forests occur in the mid-hills, where complex relations between the forests, agriculture and human subsistence exist. About 83% of the people depend on subsistence agriculture, and the forest provides food, medicine, energy, shelter, bedding materials, wood and non-wood products that support the subsistence-farming system in rural areas of Nepal. Rural people, because of their dependence on a variety of forest products, have for a long time played an important role in the use and management of the forests.

Since 1978, the main forest management strategy of Nepal is community forestry by Community Forest User Groups (CFUGs). This approach encourages active participation of local people in forest management regarding utilization and distribution of benefits from the forest. During the 1990s, community-based forest management gained momentum in Nepal, in part driven by a growing international interest in the approach to forest management. Many studies have demonstrated that community-based forest management resulted in more efficient use of forest resources as well as enhanced landscape-level forest restoration through spurred tree plantation and reclamation of forest in landside areas and river banks, which led to improvement in forest cover and reduced flash floods and associated landslides (Gautam *et al.*, 2004). Through restoration activities, between 1990 and 2010, community forests in Nepal converted sparse forests into more dense forests and non-forested areas into forested areas at rates significantly higher than in forests managed by other types of regimes (Niraula *et al.*, 2013).

The outcomes seen here are, in part, remarkable because of the devolution of decision-making authority to local user groups. Community forestry in Nepal vests rights of access, use, exclusion and management of national forestland to local groups. As the discussion in the chapter has outlined, local decision-making centres are one of three key markers of functional polycentric governance. Others are present in this case as well. For example, there appears to be heavy involvement of government foresters in crafting user groups' constitutions and operational plans (Springate-Baginski *et al.*, 2003), highlighting the linkage between local user groups and centralized governments. Similarly, part of the momentum gained in the 1990s was due to recognition, and support, from international development institutions such as the Swiss Agency for Development and Cooperation (SDC), which supported, through the Nepal Swiss Community Forestry Project, the implementation of Nepal's forestry sector plan. More recently, mechanisms have been established that link CFUGs to the World Bank's Forest Carbon Partnership Facility (FCPF) through REDD+ (reducing emissions from deforestation and forest degradation, and the role of conservation, sustainable management of forests, and enhancement of forest carbon stocks in developing countries) programmes that exchange financial payments in exchange for verifiable emissions reductions. REDD+ programmes have also catalysed efforts that coordinate CFUGs at a watershed scale and have established a watershed advisory committee that works with local-level actors and higher-level actors (multiple spatial and nested decision-making centres).

arrangements have emerged as viable governance alternatives for forest restoration. It also suggests that additional outreach and education may be required to build relationships between managers and stakeholders, who still see legitimacy conferred through top-down arrangements. If managers adhere to a technocratic discourse, that managing forests for sustained yield is the sole jurisdiction of the state, it is difficult to see the incentives for participation (Rossiter, 2008). Claims of territoriality can impinge on the success of polycentric collaboration. To maximize stakeholder buy-in, polycentric approaches should be designed with a range of perspectives in mind.

Bringing multiple types of knowledge into decision-making

The knowledge relevant for FLR will be distributed across the different actors and agencies involved in the process (see Chapter 12). Given that FLR draws on scientific, traditional and experiential knowledge of a landscape and its history, it is unlikely that a single individual or entity will have access to all this knowledge to bring it into the FLR process. Doing so will require a coordinated approach to ensure that knowledge exchange processes are legitimate and respect the contributions of different actors. Knowledge exchange is a social process that is enabled through dialogue, trust and respect between different actors (Reed *et al.*, 2014). The decentralized and collaborative decision-making units in a polycentric system provide an ideal arena in which different knowledge holders can come together to share expertise and insights and learn collectively about how to address the challenges of FLR. As these decision-making units will likely be distributed across a landscape, they provide a venue for face-to-face discussion and knowledge exchange, which is critical to provide a shared understanding across different perspectives and knowledge (after Ostrom, 2010).

Ensuring that this knowledge exchange is connected into the systems of decision-making is critical to enable adaptive management and adaptive governance of an FLR initiative. Too often, the systems that generate knowledge about a problem are separated from the arenas where decisions are made about how to deal with the problem. Adequately bringing different types of knowledge into decision-making requires mechanisms that effectively enable governance and management to respond to

Box 11.2 Habitat 141° – highlighting the mechanisms to connect nodes of decision-making across scales in Australian landscape restoration

Habitat 141° is a diverse alliance seeking to protect and restore a highly fragmented landscape. The initiative hopes to restore connectivity between national parks and nature reserves over a 700 km stretch straddling the South Australian and Victorian border into the New South Wales rangelands (see Figure 11.3). Covering 18 million hectares, Habitat 141° encompasses diverse ecological communities: heathland, mallee country, river red gum forests, flood plains, grassy woodlands and coastal ecosystems. Varied rainfall and soil quality result in a diversity of land use, from large areas of conservation, intensive forestry and cropping land use through to extensive grazing.

The alliance includes public, private and civil society actors: three state government agencies (Parks Victoria, Victorian Department of Sustainability and Environment, South Australian Department of Natural Resources), six of Australia's 56 statutory Natural Resource Management (NRM) bodies, and small and large conservation NGOs. They are united by a 50-year vision: 'To work with communities to conserve, restore and connect habitats for plants and wildlife on a landscape-scale from the outback to the ocean.'

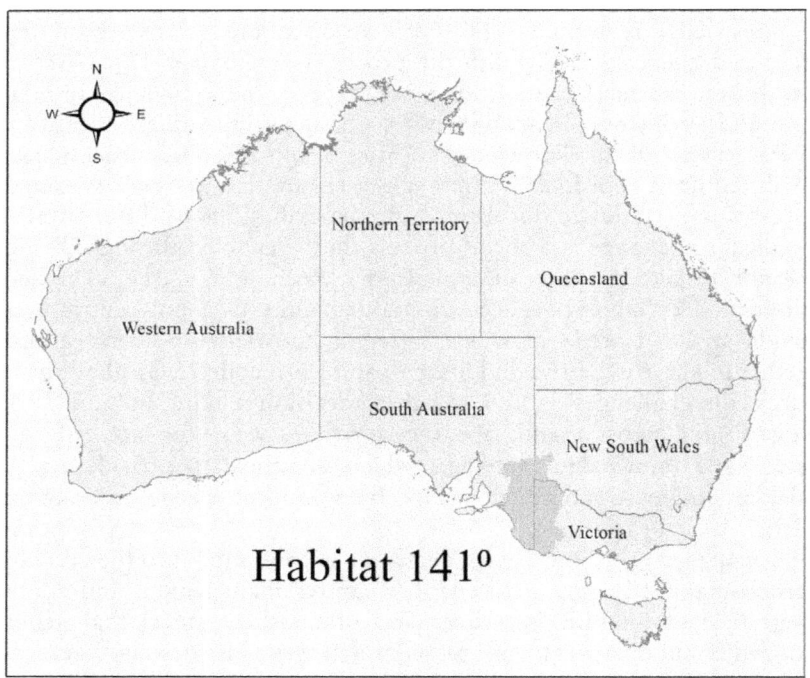

Figure 11.3 Map of the Habitat 141° spatial extent in the Australian state of Victoria.

Habitat 141°'s primary role is to facilitate collaboration between partner organizations. Discussion about Habitat 141° began in 2005, with initial conservation planning starting in 2008. Between 2009 and 2012, a 'governance working group' explored different options for how the alliance would be governed under a polycentric approach. The group envisaged a governance structure nested across four scales: a coordinating entity; collaborative planning units or 'zones'; individual projects (to be undertaken collectively or by project partners independently); and internal governance of partner organizations. Figure 11.4 depicts an idealized image of the network, where planning units, projects and partners align with the overarching vision. Decision-making was to be distributed based on the principle of subsidiarity, whereby decisions were devolved to the lowest level possible to conduct the task.

The coordinating entity comprised a council and an executive to oversee governance of the whole initiative. The council was to be a collaboration of partner organizations with a smaller executive of between five and seven elected members of the council (with provisions to bring in necessary skills). The coordinating entity was to support collaboration, undertake strategic planning, raise funds, and promote the vision and the initiative. Habitat 141° is broken into nine zones, loosely delineated according to the biophysical and social fabric of the region: the ecology, management agencies and community networks. Open to partner organizations, the zones undertake conservation action planning (CAP) (see Groves *et al.*, 2002), engage local communities, and develop and implement projects. The zones have minimal governance rules, and participation is voluntary to provide flexibility for partners to engage at their discretion. Technical working groups (for example, a governance working group, a science working group and a spatial analysis working group) would form on an ad hoc basis to provide additional expertise where necessary.

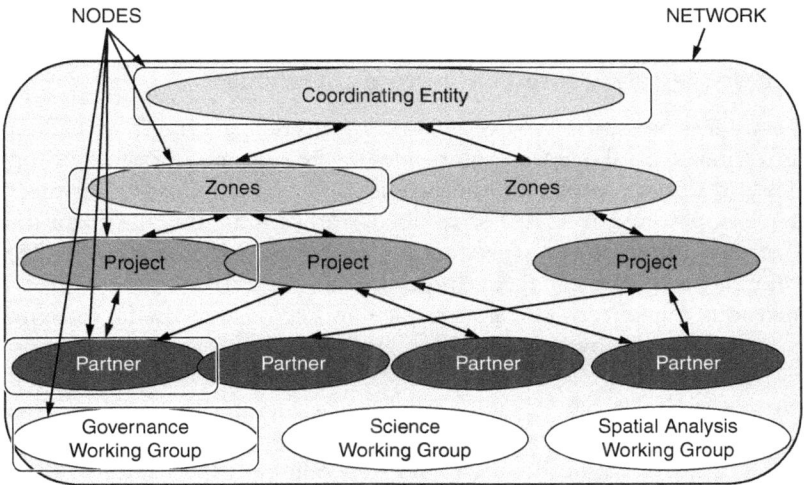

Figure 11.4 The envisaged governance network of Habitat 141°.

Source: Wyborn (2015b), reproduced with permission.

In its basic form, the governance envisaged for Habitat 141° was constructed to enable a diverse group of organizations distributed across a large landscape to coordinate and collaborate on restoration activities. It should be noted that the group faced challenges in finding sufficient resources to fund the entire entity as envisaged, so a smaller subset of activities and collaborations was undertaken, based on the broad vision and polycentric governance structure that were established in 2015.

advances in knowledge or insight about a problem (Wyborn, 2015a). Again, a polycentric approach can facilitate such connections, as the decision-making units themselves can become the arenas of knowledge exchange and learning. Studies have suggested that monitoring of forest conditions is best conducted through a coordinated approach that includes local community-level efforts as well as national and regional government inputs (Nagendra and Ostrom, 2012). Connecting monitoring data that has been collected by actors at different scales into decision-making at different scales can be facilitated by a polycentric approach, as it is likely easier to connect decision-making units with data collected at the relevant scales of monitoring and governance. It should be noted, however, that a coordinated system of monitoring, management and governance requires sufficient resources to build such connections between these different elements of a governance system, without which the capacity of a polycentric network, as a whole, to learn, adapt and evolve can be limited (Wyborn, 2015b). Thus, we suggest that a polycentric approach can enable a collective body of knowledge to be ready to deploy at the right scale when it is needed.

Synthesis: restoring forests to restore governance

In many instances, the type of restoration needed intersects with the type of governance available (see Figure 11.5). The right mix of ecological principles and science required to operate is somewhat similar for various projects (e.g. planting trees vs. streambed restoration), but the institutional players are different (e.g. for a degraded political landscape, NGOs and development agencies may be more crucial players). We map these two dimensions out in terms of governance and ecosystem needs, recognizing that sometimes a governance system may be more in need of restoration than the forest itself.

In the case of a breakdown of political structures, a focus on governance may be needed before restoration activities can take place. The type of forest restoration needed will vary depending on the health of the ecosystem. In turn, the level of input required from residents, communities, scientific managers and governing bodies may necessitate a deeper consideration of the strategies employed. Additionally, it may be worth asking

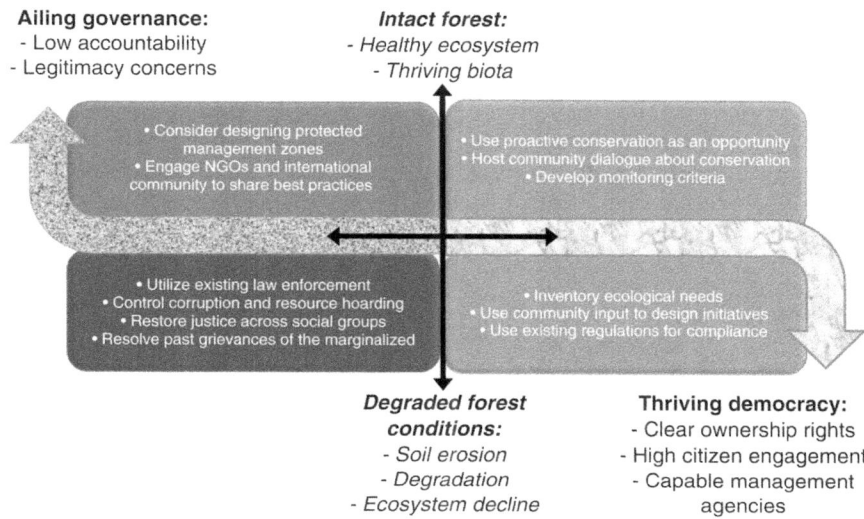

Ailing governance:
- Low accountability
- Legitimacy concerns

Intact forest:
- Healthy ecosystem
- Thriving biota

• Consider designing protected
management zones
• Engage NGOs and international
community to share best practices

• Use proactive conservation as an opportunity
• Host community dialogue about conservation
• Develop monitoring criteria

• Utilize existing law enforcement
• Control corruption and resource hoarding
• Restore justice across social groups
• Resolve past grievances of the marginalized

• Inventory ecological needs
• Use community input to design initiatives
• Use existing regulations for compliance

Degraded forest
conditions:
- Soil erosion
- Degradation
- Ecosystem decline

Thriving democracy:
- Clear ownership rights
- High citizen engagement
- Capable management
agencies

Figure 11.5 Dimensionality of Polycentric Forest Restoration: Governance and Forest Needs and Recommendations. Mapping the dimensions of governance and forest condition reveals different tactics and strategies for restoration, considering citizen needs, existing governance processes, and overall forest condition.

whether the physical act of restoration will be part of the process of restoring those political systems. In this case, is the process of inclusion through silvicultural and forest management interventions sufficient to bring governance systems back to health?

Conclusion

There are many instances where landscape goals and restoration needs require governance arrangements that are capable of yielding to new science, information and the reflexive engagement of local communities and central planners. These situations are not unique to developing countries, advanced economies, cultural perspectives, or even forest type and condition. Across various landscapes, governance arrangements can benefit from inclusion of multiple actors and knowledge types. Polycentric restoration offers the needed flexibility to encompass multiple objectives under a variety of frameworks. Building from a comprehensive awareness of local concerns, and incorporating the realities of funding partners and governmental agencies, polycentricity requires ongoing engagement from all parties. An emphasis on inclusion from input through to design, implementation, monitoring and evaluation will ensure that local values and the most current scientific understanding are maintained, along with democratic principles of engagement. Polycentric restoration encompasses planning and monitoring, but is also action-orientated. A polycentric approach enables multiple actors to have a voice, and then sets up nested decision-making scales that contribute to the ongoing decision-making support system. The ideal polycentric governance system provides the framework to coordinate activities, conduct monitoring with multiple inputs, and conduct restoration work. It represents one, but not the only, mode of governance for restoration. The process helps to identify needs and match them with the resources to accomplish goals.

References

Agrawal, A. and Ostrom, E. (2001), 'Collective action, property rights, and decentralization in resource use in India and Nepal', *Politics & Society*, Vol. 29 No. 4, pp. 485–514.

Andersson, K.P. and Ostrom, E. (2008), 'Analyzing decentralized resource regimes from a polycentric perspective', *Policy Sciences*, Vol. 41 No. 1, pp. 71–93.

Armitage, D., Berkes, F. and Doubleday, N. (2007), 'Moving beyond co-management', in Armitage, D., Berkes, F. and Doubleday, N. (eds), *Adaptive Co-Management: Collaboration, Learning, and Multi-Level Governance*, University of Washington Press for UBC Press, Seattle.

Avant, D., Finnemore, M. and Sell, S. (2010), *Who Governs the Globe?*, Cambridge University Press, Cambridge, UK.

Bentley, T., Olapade, M., Wambua, P. and Charron, N. (2015), *Afrobarometer Round 6: Where to Start? Aligning Sustainable Development Goals with Citizen Priorities: Dispatch No. 67*, available at: http://afrobarometer.org/sites/default/files/publications/Dispatches/ab_r6_dispatchno67_african_priorities_en.pdf (accessed 28 June 2017).

Berkes, F. (2010), 'Devolution of environment and resources governance: trends and future', *Environmental Conservation*, Vol. 37 No. 4, pp. 489–500.

Berkes, F.J., Colding, J. and Folke, C. (2003), *Navigating Social-Ecological Systems: Building Resilience for Complexity and Change*, Cambridge University Press, Cambridge, UK.

Bixler, R.P. (2014), 'From community forest management to polycentric governance: assessing evidence from the bottom-up', *Society & Natural Resources*, Vol. 27 No. 2, pp. 155–169.

Bixler, R.P., Johnson, S., Emerson, K., Nabatchi, T., Reuling, M., Curtin, C., Romolini, M. and Grove, J.M. (2016), 'Networks and landscapes: a framework for setting goals and evaluating performance at the large landscape scale', *Frontiers in Ecology and the Environment*, Vol. 14 No. 3, pp. 145–153.

Bixler, R.P. Reuling, M., Johnson, J. and Tabor, G. (2018), 'The Crown of the Continent: a case-study of collaborative climate adaptation', *Encyclopedia of the Anthropocene*, Vol. 2, pp. 307–315.

Bodin, Ö. and Prell, C. (2011), *Social Networks and Natural Resource Management: Uncovering the Social Fabric of Environmental Governance*, Cambridge University Press, Cambridge, UK.

Cadman, T. (2011), *Quality and Legitimacy of Global Governance: Case Lessons from Forestry*, Palgrave Macmillan, Basingstoke, England.

Cupps, D.S. (1977), 'Emerging problems of citizen participation', *Public Adminis-tration Review*, Vol. 37 No. 5, pp. 478–487.

DellaSala, D.A., Martin, A., Spivak, R., Schulke, T., Bird, B., Criley, M., van Daalen, C., Kreilick, J., Brown, R. and Aplet, G. (2003), 'A citizen's call for eco-logical forest restoration principles and criteria', *Ecological Restoration*, Vol. 21 No. 1, p. 14.

Derak, M., Cortina, J., Taiqui, L. and Aledo, A. (2017), 'A proposed framework for participatory forest restoration in semiarid areas of North Africa', *Restor-ation Ecology*, Vol. 26 No. S1, pp. S18–S25.

Dietz, T., Ostrom, E. and Stern, P. (2003), 'The struggle to govern the commons', *Science*, Vol. 302, p. 1907.

Folke, C., Hahn, T. and Olsson, P. (2005), 'Adaptive governance of social-ecological systems', *Annual Review of Environment and Resources*, Vol. 30, pp. 441–473.

Gautam, A.P., Shivakoti, G.P. and Webb, E.L. (2004), 'Forest cover change, physi-ography, local economy, and institutions in a mountain watershed in Nepal', *Environmental Management*, Vol. 33 No. 1, pp. 48–61.

Gerlak, A.K. and Heikkila, T. (2006), 'Comparing collaborative mechanisms in large-scale ecosystem governance', *Natural Resources Journal*, Vol. 46 No. 3, pp. 657–707.

Gratzfeld, J. and Khan, A.U. (2015), *Dry Woodlands in Pakistan's Punjab Prov-ince – Piloting Restoration of Unique yet Vanishing Natural Assets*, Richmond, UK, available at: www.bgci.org/files/Worldwide/News/2015jan_feb/Pakistan-Woodlowres.pdf (accessed 28 June 2017).

Groves, C.R., Jensen, D.B., Valutis, L.L., Redford, K.H., Shaffer, M.L., Scott, J.M., Baumgartner, J.V., Higgins, J.V., Beck, M.W. and Anderson, M.G. (2002), 'Plan-ning for biodiversity conservation: putting conservation science into practice', *BioScience*, Vol. 52 No. 6, pp. 499–512.

Gulbrandsen, L.H. (2012), 'International forest politics: intergovernmental failure, non-governmental success?', in Andresen, S., Boasson, E.L. and Honneland, G. (eds), *International Environmental Agreements: An Introduction*, Routledge, New York and London, pp. 151–170.

Hosonuma, N., Herold, M., Sy, V. De, Fries, R.S. De, Brockhaus, M., Verchot, L., Angelsen, A. and Romijn, C. (2012), 'An assessment of deforestation and forest degradation drivers in developing countries', *Environmental Research Letters*, Vol. 7 No. 4, p. 44009.

Humphreys, D. (1999), 'The evolving forests regime', *Global Environmental Change*, Vol. 9, pp. 251–254.

INTERPOL. (2016), *Uncovering the Risks of Corruption in the Forestry Sector: Project LEAF (Law Enforcement Assistance for Forests)*, Lyon, France, available at: www.google.com/url?sa=t&rct=j&q=&esrc=s&source=web&cd=1&ved=0ahU KEwjvqYP689_YAhVDLKwKHUFuD14QFggpMAA&url=https%3A%2F%2F www.interpol.int%2FMedia%2FFiles%2FCrime-areas%2FEnvironmental-crime %2FUncovering-the-Risks-of-Corruption-in-the-Forestry-Sector (accessed 28 June 2017).

Jacobson, C. and Robertson, A.L. (2012), 'Landscape conservation cooperatives: bridging entities to facilitate adaptive co-governance of social–ecological systems', *Human Dimensions of Wildlife*, Vol. 17 No. 5, pp. 333–343.

Jedd, T. and Bixler, R.P. (2015), 'Accountability in networked governance: learning from a case of landscape-scale forest conservation', *Environmental Policy and Governance*, Vol. 25 No. 3, pp. 172–187.

Kahler, M. (2009), *Networked Politics: Agency, Power, and Governance*, Cornell University Press, Ithaca.

Lau, M. (2014), 'Flexibility with a purpose: constructing the legitimacy of spatial governance partnerships', *Urban Studies*, Vol. 51 No. 9, pp. 1943–1959.

Lemos, M.C. and Agrawal, A. (2006), 'Environmental governance', *Annual Review of Environment and Resources*, Vol. 31 No. 1, pp. 297–325.

Mansourian, S. (2016), 'Understanding the relationship between governance and forest landscape restoration', *Conservation and Society*, Vol. 14 No. 3, pp. 267–278.

Mansourian, S. (2017), 'Governance and forest landscape restoration: a framework to support decision-making', *Journal for Nature Conservation*, Vol. 37, pp. 21–30.

Nagendra, H. and Ostrom, E. (2012), 'Polycentric governance of multi-dimensional forested landscapes', *International Journal of the Commons*, Vol. 6 No. 2, pp. 104–133.

Newell, P.J., Pattberg, P. and Schroeder, H. (2012), 'Multiactor governance and the environment', *Annual Review of Environment and Resources*, Vol. 37, pp. 365–387.

Newman, J. (2004), 'Constructing accountability: network governance and managerial agency', *Public Policy and Administration*, Vol. 19 No. 4, pp. 17–33.

Nie, M. and Metcalf, P. (2015), *The Contested Use of Collaboration and Litigation in National Forest Management*, Bolle Center for People and Forests, Missoula, MT.

Niraula, R.R., Gilani, H., Pokharel, B.K. and Oamer, F.M. (2013), 'Measuring impacts of community forestry program through repeat photography and satellite remote sensing in the Dolakha district of Nepal', *Journal of Environmental Management*, Vol. 126 No. 15, pp. 20–29.

Orchard, S.E. and Stringer, L.C. (2016), 'Challenges to polycentric governance of an international development project tackling land degradation in Swaziland', *Ambio*, Vol. 45 No. 7, pp. 796–807.

Ostrom, E. (1999), 'Coping with tragedies of the commons', *Annual Review of Political Science*, Vol. 2 No. 1, pp. 493–535.

Ostrom, E. (2010), 'Polycentric systems for coping with collective action and global environmental change', *Global Environmental Change*, Vol. 20 No. 4, pp. 550–557.

Ostrom, V., Tiebout, C.M. and Warren, R. (1961), 'The organization of government in metropolitan areas: a theoretical inquiry', *American Political Science Review*, Vol. 55 No. 4, pp. 831–842.

Pedynowski, D. (2003), 'Prospects for ecosystem management in the Crown of the Continent ecosystem, Canada-United States: survey and recommendations', *Conservation Biology*, Vol. 17 No. 5, pp. 1261–1269.

Reed, M.S., Stringer, L.C., Fazey, I., Evely, A.C. and Kruijsen, J.H.J. (2014), 'Five principles for the practice of knowledge exchange in environmental management', *Journal of Environmental Management*, Vol. 146, pp. 337–345.

Rossiter, D.A. (2008), 'Producing provincial space: crown forests, the state and territorial control in British Columbia', *Space and Polity*, Vol. 12 No. 2, pp. 215–230.

Schilling, J., Freier, K.P., Hertig, E. and Scheffran, J. (2012). 'Climate change, vulnerability and adaptation in North Africa with focus on Morocco', *Agriculture, Ecosystems, and Environment*, Vol. 156, pp. 12–26.

Schultz, C.A., Jedd, T. and Beam, R.D. (2012), 'The Collaborative Forest Landscape Restoration program: a history and overview of the first projects', *Journal of Forestry*, Vol. 110 No. 7, pp. 381–391.

Scott, J.C. (1998), *Seeing Like a State: How Certain Schemes to Improve the Human Condition Have Failed*, Yale University Press, New Haven, CT.

Springate-Baginski, O., Dev, O.P., Yadav, N.P. and Soussan, J. (2003), 'Community forest management in the middle hills of Nepal: the changing context', *Journal of Forest and Livelihood*, Vol. 3 No. 1, pp. 5–20.

Thomas, D.S. and Twyman, C. (2005), 'Equity and justice in climate change adaptation amongst natural-resource-dependent societies', *Global Environmental Change*, Vol. 15 No. 2, pp. 115–124.

Tyrväinen, L., Silvennoinen, H., Nousiainen, I. and Tahvanainen, L. (2001), 'Rural tourism in Finland: tourists' expectation of landscape and environment', *Scandinavian Journal of Hospitality and Tourism*, Vol. 1 No. 2, pp. 133–149.

Westerling, A.L. (2016), 'Increasing western US forest wildfire activity: sensitivity to changes in the timing of spring', *Philosophical Transactions of the Royal Society B: Biological Sciences*, Vol. 371 No. 1696, available at: http://rstb.royalsocietypublishing.org/content/371/1696/20150178.abstract (accessed 28 June 2017).

World Resources Institute. (2013), 'Assessing forest governance', *The Governance of Forests Initiative Indicator Framework*, available at: www.wri.org/publication/assessing-forest-governance (accessed 28 June 2017).

Wyborn, C. (2015a), 'Co-productive governance: a relational framework for adaptive governance', *Global Environmental Change*, Vol. 30, pp. 56–67.

Wyborn, C. (2015b), 'Cross-scale linkages in connectivity conservation: adaptive governance challenges in spatially distributed networks', *Environmental Policy and Governance*, Vol. 25 No. 1, pp. 1–15.

Wyborn, C. and Bixler, R.P. (2013), 'Collaboration and nested governance: scale dependency, scale framing, and cross-scale interactions in collaborative conservation', *Journal of Environmental Management*, Vol. 123 No. 15, pp. 58–67.

12 Integration of Traditional and Western knowledge in forest landscape restoration

Frank K. Lake, John Parrotta, Christian P. Giardina, Iain Davidson-Hunt and Yadav Uprety

Introduction

Given the diversity of environmental, historical, social, economic and cultural contexts in which forest landscapes (and their degradation) occur, restoration efforts require the effective engagement and mobilization of the social and human capital that exists within these socio-ecological systems. This involves the utilization of diverse knowledge systems (von der Porten and de Loë, 2014), which in some cases, includes a significant body of Traditional knowledge and expertise that was marginalized during colonial periods (Stewart, 2002). Importantly, these knowledge systems remain largely misunderstood and underappreciated by contemporary, Western-trained managers, planners, resource specialists, foresters, scientists and other professionals (Trosper *et al.*, 2012b). It can be argued, however, that there is a growing recognition of the importance of incorporating information from Traditional knowledge systems, alongside Western knowledge, into land management decision-making, especially where there is the goal of better aligning management with the varied and growing needs and place-based objectives of diverse stakeholders (Klooster, 2002; Charnley *et al.*, 2007; Uprety *et al.*, 2012; Chazdon *et al.*, 2017; Sterling *et al.*, 2017; Diaz *et al.*, 2018) and the historical ecological realities of forest landscapes.

It is our contention that well-integrated approaches to forest landscape restoration (FLR) will be most effective at mitigating the numerous, often persistent drivers of landscape degradation. Parties involved in formulating a restoration strategy need to reach a common understanding of 'what degradation means and is'. Conflicts can arise between indigenous/local communities and other stakeholders who see very differently the factors that have degraded and continue to degrade not only the landscape but also ecological processes and socio-cultural practices. Lack of consensus among all FLR parties on the applications of Traditional and local knowledge to managed resources can undermine top-down, often industrial/commercial focused, efforts to manage agricultural lands, rangelands, forests and other ecosystems that have valued resources and services.

By engaging broader approaches to FLR, including the integration of multiple knowledge systems, we suggest that local communities, agencies, public user groups, and downstream or adjacent ownerships will be better able to participate in or even lead the management process, to the benefit of both natural and human systems (Berkes, 2009; Gilmour, 2016; Sterling *et al.*, 2017). While desirable, shifting from a model of government agency-driven, top-down management to a model of shared governance can represent a significant challenge (Brown, 2005; Gilmour, 2016). As discussed in earlier chapters of this book, achieving the aims of FLR – defined here as returning ecological integrity and resilience to deforested or degraded landscapes while enhancing human wellbeing – requires recognition that landscapes are not simple biophysical spaces, but rather, complex socio-ecological systems that simultaneously shape and are shaped by people (Berkes, 2009; Gilmour, 2016). Further, landscape-scale, community-based and intergenerational stewardship of place has sustained residents within and beneficiaries outside of landscape boundaries for millennia (Berkes *et al.*, 2009). These realities are often dismissed in depictions of landscapes as once pristine ecological systems that have been degraded by contemporary user groups (Botkin, 2004).

In a highly integrated, relationship-driven framework for FLR, collaborative management creates a pathway for negotiating and reconciling the needs and objectives of diverse stakeholders (Brown, 2005; Berkes, 2007; Gilmour, 2016). In the case of community-based conservation, for example, Berkes (2007) notes that

> improving the integration of conservation and development requires rethinking conservation by using a complexity perspective and the ability to deal with multiple objectives, use of partnerships and deliberative processes, and learning from commons research to develop diagnostic tools. Perceived this way, community-based conservation has a role to play in a broad pluralistic approach to biodiversity protection: it is governance that starts from the ground up and involves networks and linkages across various levels of organization.

In this chapter, we will explore the process of integrating multiple knowledge systems in FLR, with a primary focus on experiences in North America (see also Trosper *et al.*, 2012a). We will first examine the values and worldviews underlying Western and Traditional knowledge systems, as well as specific efforts to integrate knowledge systems in restoration and management at landscape scales in a variety of ecological and socio-cultural contexts. We consider the challenges – and opportunities – involved with integrating Traditional and Western knowledge systems at various stages of the FLR planning and implementation process, and how these challenges have been met, through case studies in Canada and the US. Insights gained from these analyses are summarized in an FLR planning cycle model, which seeks to accommodate Traditional knowledge to a greater extent than those previously considered (e.g. by Vallauri *et al.*, 2005).

Traditional peoples, stakeholders and FLR managers will learn of each other's interests, restoration goals and values associated with resources or specific habitats of the landscape. Different forms of knowledge reflect various ways of understanding ecological processes, socio-cultural stewardship practices and management systems. Different knowledge systems may also have strengths and weaknesses at different scales of social or ecological factors influencing the processes of FLR.

Utilizing Traditional knowledge in forest landscape restoration

Working together, FLR entities strive to understand each other's histories, potential bias, perspectives for applicable strategies, and position or role. In this section, we identify ways of recognizing historical legacies of landscapes and human communities, and consider the degradation of socio-ecological systems and approaches for respecting and reconciling worldviews and interests by examining model approaches. Each section addresses methods that integrate Traditional knowledge and indigenous stewardship practices with Western knowledge and management practices in FLR strategies.

Recognizing historical legacies, understanding degradation, and respecting cultural identity and diverse worldviews

An important starting point for collaborative management towards FLR objectives, specifically those involving indigenous communities, is an appreciation that (a) their Traditional knowledge belongs to their communities and (b) these communities continue to carefully maintain and thoughtfully articulate their ancestral connections to place. These connections to place are complex and rich despite centuries of colonial subjugation and contemporary efforts to disenfranchise indigenous peoples of their lands, resources and culture in most parts of the world (Parrotta and Trosper, 2012). It is important to recognize that, in many parts of the world, natural resources were a central focus of colonizing and/or modern state-driven forces, which brought with them approaches to resource exploitation and management that favoured centralized decision-making and methods that often conflicted (and continue to conflict) with the community-based decision-making and resource management strategies of indigenous communities.

The inherited colonial history of many North American landscapes, and other landscapes globally, left not only a biophysical legacy but also a perceptual legacy that (1) emphasizes centralized control of decision-making, planning, resource management, monitoring and communication and (2) minimizes the role of indigenous communities, whose involvement is often restricted to serving in volunteer programmes and/or providing feedback on proposed activities through limited avenues of state government to tribal government consultation, and communication during narrow public comment periods. As a result, colonial histories continue to shape interactions among governmental land management agencies and communities (Williams, 1990), with forest industry and conservation enterprises often defining how FLR is conceived, implemented and assessed, and how the actual or expected economic and other benefits of FLR are allocated among 'stakeholders'.

Such important historical legacies need to be addressed at the outset when government agencies, or others with jurisdictional control over lands, seek meaningful involvement of local and, particularly, indigenous communities in FLR planning and management. As von der Porten and de Loë (2014) point out, indigenous communities perceive their role in collaborative environmental governance to be one of ancestrally defined, sovereign nations versus the 'stakeholder' roles assumed by those seeking knowledge from indigenous communities. Within this context, it may be difficult for Western-trained professionals to see traditional knowledge systems as pillars of cultural identity that support indigenous peoples' efforts to maintain (and in some cases, restore) their place as nations situated on (often illegally) occupied 'settled' landscapes. For those seeking Traditional knowledge to help inform FLR, historical legacies that continue to negatively impact indigenous communities are often seen as part of a dark but distant past that has at least partially been resolved, for example through negotiated treaty and legal processes (von der Porten and de Loë, 2014). Indigenous cultural lifeways and practices, as well as personal, familial and community wellbeing, all depend on intimate relationships with the natural world (see Table 3.1) (Donatuto *et al.*, 2014; Kealiikanakaoleohaililani and Giardina, 2016). The task of integrating Traditional knowledge systems into an FLR process, therefore, requires an openness to conversations that challenge assumptions about Western contemporary knowledge of history, governance, stewardship, and the cultural linkages between indigenous people and their ancestral lands (Davidson-Hunt and O'Flaherty, 2007).

Most socio-ecological landscapes continue to be impacted by colonial legacies (e.g. overlapping tenure and property rights, land use practices, fire suppression and exclusion, industrial forest management, top-down decision-making approaches and structures), and some lands requiring FLR underwent often violent removals of ancestral residents (e.g. indigenous peoples and local communities). For this reason, we argue that contemporary professionals leading efforts to integrate Traditional knowledge into FLR should acknowledge, examine, and seek to better understand colonial features and other historical legacies within current restoration approaches (cf. Hall and Fenelon, 2016). This process of understanding assumptions, preconceptions and biases is required to effectively build respectful, long-term relationships that lead to partnerships that eventually can support integrated FLR planning and implementation. It should be noted that these recommendations are complementary to, and could be relied on to support, other models proposed for FLR planning (e.g. Dudley *et al.*, 2005; Vallauri *et al.*, 2005; Giardina *et al.*, 2007).

Other important questions relate to who the beneficiaries of FLR will be, and over what time scale benefits are to be expected (Boedhihartono and Sayer, 2012). FLR partners work to gain recognition of the contributing factors and aim to have a shared common understanding of the nature, manifestations and root causes of degradation to socio-ecological systems. Identification of current challenges, and direct and indirect threats and stressors to the recovery of landscapes, allows the partners to formulate solutions and take actions (i.e. restoration strategies) to alleviate or effectively address agents of degradation (O'Connor *et al.*, 2005).

Such dialogue is critically important at the outset of any FLR programme for a number of reasons. Because restoration is defined as bringing about recovery to a degraded landscape, value-laden terms such as 'degraded' – whose meaning also shifts over time – can be highly problematic, particularly when dealing with indigenous landscapes that have been dramatically altered. How a landscape geographer, an ecologist or an indigenous practitioner views 'degradation' will be defined by their particular biophysical or environmental value system, operational context, epistemological framework, and history with the landscape. Approaches to incorporating different knowledge systems into FLR, in part, have to question underlying assumptions about the role humans have as 'degraders' or active agents of ecosystem recovery across different landscapes. Regardless of which partnership model is involved in FLR, complicated and often difficult questions need to be addressed so that the partnership members develop both a common understanding of what degradation means and a shared understanding of the causes. Further, since degradation is not only ecological but inclusive of socio-cultural systems, there may be differences in perspective that need to be discussed regarding which parts of the coupled human and natural system are degraded, and whether an intact ecological system can be considered degraded because it does, or cannot for legal reasons (e.g. endangered species conservation), sustain the human members of the larger community (Andrade and Rhodes, 2012).

Few Western-trained ecologists appreciate the extent and depth to which indigenous peoples throughout the world have gone to manage and so transform landscapes prior to European colonization (Mann, 2002; Stewart, 2002; Bowman *et al.*, 2011). This led early ecologists and anthropologists to suggest that pre-colonial landscapes were relatively 'pristine', a conservation belief deeply embedded in Western knowledge systems (Nash, 2014) that is challenged in the Americas by overwhelming evidence about the de-population of these landscapes following European contact with indigenous peoples and subsequent changes to those landscapes (Cronon, 1996; Wallman *et al.*, 2018; see Box 12.1). As a result, the question of

Box 12.1   The role of indigenous fire management practices in shaping landscapes and their ecologies in North America

Mythology about historical wilderness areas free of human influence has long dominated conservation discussions, while the rich history of indigenous management across most of North America, and other areas globally, has been largely ignored (Nash, 2014). A poignant example of this divide that can separate knowledge systems is the tall grass prairies in the central plains of North America and oak savannahs in the western United States. These landscapes are commonly thought of as having been maintained primarily by natural fires (e.g. lightning), when in fact they were created and maintained though active indigenous burning (Stewart, 2002). If not for millennia of prescribed burning by Native Americans seeking to create rangelands for the American bison and other early seral dependent game species, a practice actively continued by homesteading ranchers to the present day, much of North America's tall grass prairie and savannah would revert to forest. These tall grass prairie and savannah systems have evolved to include a remarkable diversity of fire-adapted grass, forb and some tree species that in turn shape the ecosystem and characterize that landscape, but these systems are disappearing because natural fires (e.g. lightning ignitions) are too infrequent and rainfall too high to maintain fire-adapted species. This loss to forest recovery, ironically, is alarming to ecologists who view these systems as 'natural'. Are these systems degraded forests in need of restorative fire exclusion? Or rather, should their grassy, open condition be viewed positively as the result of millennia of interactions with human communities?

In the western United States, significant investments are being made to restore landscapes in order to reduce fire danger. Ironically, nearly all of the western United States had been intensely managed with fire by native peoples until conquest (a little over a century ago) replaced indigenous management of landscapes with practices that resulted in the removal of large, commercially valuable, fire-resistant trees, which have since been replaced by dense regrowth of small-diameter trees, introduction of grazing animals that reduced very fine fuels (i.e. grasses/forbs), and landscape-scale attempts to suppress any and all fires (Stewart, 2002). This combination of practices largely explains the severe mega-fire events plaguing western US states in recent years (Hessburg *et al.*, 2015).

Many non-indigenous managers and the public, who are less dependent on fire-induced resources in this region, view a more closed, densely 'forested' landscape as beneficial for conservation, recreation and scenic values. This contrasts with the views of local indigenous communities, who consider the same landscape as not adequately burned through a functional cultural fire regime promoting 'open' forests that could provide more water, foods, materials, medicines and ecosystem services, and sustain fire-dependent cultural practices and knowledge systems in a heterogeneous and resilient landscape (Berkes and Davidson-Hunt, 2006; Hessburg *et al.*, 2015; Lake *et al.*, 2017). Consequently, the restoration of more open fire-adapted forests and the reinstatement of tribal/indigenous burning are prevented in some areas of the landscape in order to protect the currently suitable transitional habitat for endangered wildlife, a habitat condition resulting from industrial forestry operations and fire exclusion practices (Kimmerer and Lake, 2001; Long *et al.*, 2016). A philosophical difference thus arises for FLR planning and management: 'ecosystem services for humans, or humans services for ecosystems?'

which attributes of the landscape should be considered as an historical reference by restoration strategists and partners is complicated by such flawed conceptual models of landscape condition and change.

When entering into a process in which Traditional knowledge is sought for informing FLR, it is valuable to consider a culturally sensitive best practices framework (Grenier, 1998). How should one request access to Traditional knowledge? If shared, how can this knowledge be used without being appropriated in an exploitive way (i.e. free prior and informed consent (UNDRIP, 2007))? In the context of integration of Traditional knowledge into Canadian water resources management, von der Porten *et al.* (2015) make six recommendations for the practice of collaborative environmental governance that are broadly relevant to FLR efforts: (1) approach or involve indigenous peoples as self-determining nations rather than as one of many collaborative stakeholders or participants; (2) identify and engage with existing or intended environmental governance processes and assertions of self-determination by indigenous nations; (3) create opportunities for relationship-building between indigenous peoples and policy or governance practitioners; (4) choose venues and processes of decision-making that reflect Indigenous rather than Eurocentric venues and processes; (5) provide resources to Indigenous nations to improve capacity for collaboration or for policy reform decision-making; and (6) find ways to support indigenous nations in their own continued environmental decision-making and 'self-determination'.

It is possible, however, to engage Traditional knowledge on a deeper level, but this requires understanding of how the foundations of Western resource management differ from those supporting traditional (i.e. indigenous and local) resource management (Berkes *et al.*, 2000, as addressed earlier in Chapter 3). These distinctions are typically not understood or even recognized by government agency land managers (von der Porten and de Loë, 2014), in part because Western management and natural resources education rarely includes examination of the epistemological framework in which (cultural) management approaches are embedded, or the value systems and assumptions upon which they are built (Chase, 1986).

There is also a need for greater understanding of how natural resource management within coupled human and natural systems has been influenced, even driven, by historical shifts from local to centralized control over forests and their management in many landscapes (Smith, 2012;

Nash, 2014; Steen-Adams *et al.*, 2015, 2017). Legacies of management control are not trivial, as responses of forest composition, structure, function and dynamics to management can take centuries to manifest.

Utilizing effective and appropriate partnership models

Optimally, the socio-economic and biocultural context in which forest landscape restoration is to proceed defines the process for initiating discussions and planning specific projects or a comprehensive, multi-decade initiative. To this end, the chances of success will be greatly enhanced if participants, including formal management organizations with jurisdictional responsibility for identified resources and informal stewardship entities that have local, indigenous or other historical ties to a landscape, come together to dialogue as partners in the process.

Partnership models are highly diverse, but may be grouped into three broad categories: (1) the *government model*, where decisions are made completely within one or more government agencies having legal control over public lands and their management, and outside (e.g. local) opinions may or may not be sought depending on the specific situation and availability of options for public input; (2) the *collaborative model*, where decision-making power is shared across formally identified partners; and (3) the *community model*, where decision-making power resides with community members who own or have had formal or informal traditional and customary rights to the management of specific territories. These models differ in a number of important respects: (1) the extent to which different knowledge systems are – or could be – utilized in resource management; (2) distribution of authority for resource use and management; (3) distribution of expected benefits (i.e. ecosystem services) to broader society versus local communities; (4) economic capacity affected by governance stability and global to local market economies; and (5) jurisdictional control or tenure ownership of specific landscapes and resources (Table 12.1). Understanding which model is operative (or feasible) in a given landscape allows more effective integration of Traditional knowledge into the visioning, planning, overall decision-making and implementation processes.

Integration of knowledge systems under these partnership models can take several forms. Within the government model, Traditional knowledge may be treated as another source of supplemental or anecdotal information for management carried out entirely by a public agency. Where there is potential for and interest in utilizing the collaborative model to integrate Traditional knowledge into FLR, and where appropriate decision-making mechanisms are developed and agreed upon, the identified partners may need to engage in a process that enhances understanding among all concerned of the epistemological underpinning of different approaches to resource management. Partners learn of the socio-economic and biocultural context that has shaped historical landscape management as well as

Table 12.1 Partnership model types and scope of knowledge systems capacity for FLR

Model type	Knowledge system application to FLR and resource management	Decision authority for resource use and management	Scope of expected landscape benefits (i.e. ecosystem services) to broader society versus local communities	Economic capacity affected by governance stability and global to local market economies	Rights to the management of specific geography: jurisdictional control or tenure ownership of resources
Government	Western strongly favoured over local/Traditional	Centralized: government agency representatives often not living on the land	Broad, to address national-scale economic interest serving society, with some local resource interests	Higher to moderate, based on global to national capital market interest, e.g. carbon cap and trade revenue	Colonial-state controlled: national-scale policies/ authorities affect local community-scale rights
Collaborative	Both Western and local/Traditional, with the former often more prominent	Semi-centralized: mixed partner representatives, some living on land	Broad to regional, to support national economic interest and serving some local water, food and forest products interests	Moderate, based on national to regional market factors affecting resource and natural capital values	State/regional controlled: national scale with allowances for local stewardship and engagement
Community	Local/Traditional generally more prominent than Western	Decentralized: local/ traditional leadership living on land. Often hereditary or with elder councils	Regional to local, to support mostly local needs and serving some regional and fewer national resource interests	Lower to moderate, based on reduced dependency on national capital markets and greater reliance on local resources for security and livelihoods	Local/regional controlled: Usufruct and hereditary rights with traditional area or resource stewardship claims

the impacts of transitions from a historical indigenous community model of resource management to a government model. Through such a mutual learning process, participants can identify shared values and interests, establish a common language and perspective on land, resources, and management strategies, build the trust necessary for the integration of Traditional knowledge into FLR, and ideally, create a framework that guides a successful collaboration model approach into the future.

Defining partnership interests, values and resource needs

Forest landscape restoration programmes benefit from defining partners' interests, the values they attribute to the landscape, and their diverse resource needs. In this section, we identify components of socio-ecological processes in FLR that integrate Traditional knowledge and indigenous stewardship practices. We focus on how partner values influence FLR strategies, as well as conservation and restoration targets and goals with respect to specific sites, habitats and/or species. We also concentrate on the collaborative governance model (versus the government or community model) because most forest landscape governance today occurs under the government governance model, and here we seek transitions to more collaborative approaches where indigenous communities are recognized as full partners in shaping the design of ecological, cultural and economic landscapes.

Inclusive stakeholder involvement and participation in collectively identifying shared values and interests is vitally important for successful outcomes of FLR, as it acknowledges what each brings to the discussion and their history with the landscape (Brown, 2005). Reconciliation and efforts to (re)build trust through dialogue among partners are essential (von der Porten et al., 2016). All partnership entities work together to identify knowledge, resources and capacities that each can contribute. They recognize their own and each other's political, economic and ecological strengths and weaknesses as potential opportunities and barriers to achieving FLR. Legal standing, tenure and governance structures of indigenous communities are factors that affect the viability of FLR approaches. Economic capacity includes partnership development for an efficient restoration programme and securing financial resources for programme planning to initiate the implementation of FLR treatments in strategic areas across the landscape. FLR goals emphasize how partners foster recovery of coupled human and natural systems over time for the production of ecosystem goods and services provided by habitats and various resources.

Analysis of past degradation (e.g. contributing factors), active pressures (identified as challenges, or threats and stressors operating at different scales) and potential influences on the landscape is needed. Political, socio-economic, cultural or ecological processes (actual or perceived) all need to be considered in the integration of Traditional and Western knowledge systems for collaborative approaches to FLR to succeed. Given the various perspectives and interests that might be represented in a partnership, conducting a needs assessment and developing restoration targets may reveal some conflicts. Defining needs and targets so that they meet various partner values is important, and ideally, discussions during the initiation phase can lead to efficient and consensual definition of restoration goals. Stakeholders (or partners) can collaboratively define restoration and conservation targets (i.e. Open Standards for the Practice of Conservation) at various spatial scales, and constrain actions within various socio-political or jurisdictional boundaries (single ownerships, watershed partnerships, eco-regional or trans-national reserves), as scales and boundaries relate to shared values (Grantham *et al.*, 2010).

In the context of FLR planning and implementation involving local and indigenous communities and their knowledge, restoration may include re-establishment of former resource stewardship responsibility and associated practices. This includes 'eco-cultural revitalization' in the form of coupled socio-ecological restoration processes at various scales: landscapes, habitats and resources (Kimmerer, 2011; Martinez, 2014).

Indigenous perspectives of conservation emphasize managing for abundance so that a resource can sustain communities. For Western-trained conservationists, restoration may be strongly aligned with preservation of biodiversity, achieved through reconnecting forest patches, buffering protected areas or creating 'stepping stones' in the landscape. For an indigenous community steward, effective management may mean focusing on actions that promote the abundance and fitness of cultural keystone species, to be harvested, gathered or otherwise sustainably used by community members as part of an intergenerational practice of culture (Joseph and Mansourian, 2005; Sterling *et al.*, 2017). Restoring a cultural keystone species or habitat in a landscape is of high priority in order to create abundance of that resource and sustain people, their knowledge systems and cultural practices. This is illustrated in the case studies of eastern white pine restoration by the Kitcisakik Algonquin community of western Quebec (Box 12.2) and the restoration of traditional *manomin* (wild rice) management by the Wabaseemoong Independent Nations in Northwestern Ontario (Box 12.3).

The identification of protection (conservation) and management as passive or active restoration strategies, coupled with cross-scale indigenous stewardship practices, informs the FLR approach. Indigenous perspectives often consider utilization (i.e. stewardship) of resources (habitats or species) as a goal of restoration. Priority areas that require restoration reflect the shared landscape values and habitats or resources most critical to perpetuating socio-ecological processes to achieve conservation goals or targets (Uribe *et al.*, 2014). Strategies may identify which landscapes, jurisdictional units or functions (ecosystem services) are to be maintained, enhanced or restored, as well as focal or indicator species (i.e. cultural

Box 12.2 White pine restoration in cultural landscape, western Quebec, Canada

The Kitcisakik Algonquin community of western Quebec desires restoration and sustainable management of eastern white pine (*Pinus strobus* L.) on its ancestral landscape, where it was abundantly available in the past. A cultural keystone tree species in the forests of eastern North America, white pine provides numerous ecosystem goods and services to indigenous peoples and has long been an important component of traditional life within the Kitcisakik Algonquin community (Uprety *et al.*, 2013). The community suggested that mixed plantations should be used in a culturally adapted restoration strategy.

The community's interest in restoration was communicated to the Université du Québec en Abitibi-Témiscamingue through the Kitcisakik Forest Management Committee. The university designed the research together with the community. Before suggesting culturally adapted white pine restoration and management strategies, it was crucial to document why and how the species was important to the community and how it could be restored. The aboriginal community's bottom-up approach called for restoration of white pine, and the research was designed by the academic institution. Restoration and management strategies integrated both Traditional and Western knowledge.

Following these findings, an ecological study was carried out in collaboration with community members to find out whether the landscape was still suitable for the restoration of white pine. Based on analysis of ecological types and sampling, it was concluded that the landscape held potential for white pine restoration, addressing their needs (Uprety *et al.*, 2014). The study illustrated that different management strategies should be used near northern range limits, where effects of site conditions and disturbance agents are different than in the centre of a species' range. Five scenarios were suggested, taking into account community needs and ecological types (potential vegetation and abiotic conditions) (Uprety *et al.*, 2017). The strategies have been endorsed by Kitcisakik Forest Management Committee, and the committee is awaiting financial support to begin the restoration plan.

This case study illustrates how the concerns of aboriginal people can be addressed by integrating Traditional and Western knowledge (see also Steen-Adams *et al.*, 2011). The aim is not to restore white pine everywhere on the territory for industrial purposes, but rather, to maintain or increase the provision of white pine-associated ecosystem services to the Kitcisakik community. The responsibility for some of the restoration and management operations on family hunting grounds could be given to community members by agreement with the Crown once guidelines and training have been provided. Such community-based approaches have been shown to be efficient, to have increased legitimacy and to be more sustainable (Ribot *et al.*, 2006).

Indigenous peoples' participation, and the recognition and inclusion of their knowledge into restoration and management projects, can contribute to building a strong partnership for successful implementation that significantly improves the social acceptability, economic feasibility and ecological viability of restoration projects (Garibaldi and Turner, 2004; Higgs, 2005; Uprety *et al.*, 2012). Therefore, a shift from 'just another stakeholder' to 'shared decision makers' (Stevenson and Webb, 2003) is possible. The approach presented here, where restoration and management scenarios take into account cultural needs and ecological constraints, could find wide application in diverse forest settings, as it could help meet the objectives of certification standards (e.g. Forest Stewardship Council (FSC)) with regard to the rights and needs of indigenous people (Uprety *et al.*, 2017).

Figure 12.1 White pine restoration in the Kitcisakik ancestral territory, western Quebec.

Notes
Top: (a) Undertaking ecological study in the landscape. (Photo: Y. Uprety.)
Bottom: (b) Young white pine forest in the cultural landscape. (Photo: Y. Uprety.)

Box 12.3 Biocultural design: harvesting Manomin (wild rice) with Wabaseemoong independent nations in northwestern Ontario, Canada

Restoration as a practice has suffered from the same challenge as ecology in general through the removal of humans as integral actors within our conceptualization of an ecological system (Davidson-Hunt and Berkes, 2003). From this perspective, restoration is often conceived as a return to the pre-settlement, or historical, condition of an ecosystem, repairing the harm done by human presence. In contrast, Wabaseemoong Independent Nations in northwestern Ontario, Canada, offer an Indigenous understanding of 'restoring' manomin (wild rice). This landscape is a mosaic of forests and lake-wetland systems in which wild rice was historically a dominant riparian/aquatic habitat. Many factors have contributed to the reduction of this culturally important species across its historical range (Pillsbury and McGuire, 2009).

Manomin (*Zizania palustris* L.), an aquatic grain, has been central to Anishinaabe economic, political, social and cultural life since time immemorial (Vennum, 1988; Davidson-Hunt *et al.*, 2005). An important source of food for Aboriginal communities, it also became a key staple during the fur trade era and a specialty product exported around the world up to the present time. It required both care for its sustained production and political negotiation among Anishinaabe communities, and other Aboriginal neighbours, for key lakes where it was harvested as well as emerging colonial governments. Harvest camps are recorded both in the ethnohistorical record and in Elders' memories as times when people, who lived dispersed during much of the year, came together and worked together to harvest rice, but also conducted other social affairs. Its centrality to Anishinaabe life was reflected in the name of the moon that occurs during its harvest, and its presence in ceremonies.

For Wabaseemoong people, such interwoven relations with manomin continued through the early post-war period of the twentieth century in spite of the fur trade and the establishment of the institutions of colonial governance in northwestern Ontario. However, increasing disruptions during the twentieth century eventually broke the weave of the manomin fabric of relations (Kuzivanova, 2016). These disruptions began in the early twentieth century with the establishment of reserves on which the Wabaseemoong were required to live to receive Treaty annuities and continued with the establishment of residential schools in the mid-century. The latter introduced children to new foods and language and separated them from their parents and land-based practices such as manomin harvesting. In the 1950s, two dams were established, one upstream and one downstream of the community. These dams disturbed the natural flow patterns of the rivers as well as flooding many wetlands and lakes attached to the rivers, destroying many manomin fields. The Wabaseemoong, along with other communities whose fields had likewise been impacted by other dams, began to harvest in eastern Manitoba, but this practice was halted by the establishment of a provincial park. Mercury contamination of the river systems led to long-term human health impacts through consumption of fish and other aquatic foods. By 2013, with the cumulative degradation of the river ecosystems and cultures, only Elders who had themselves participated in the harvest of manomin retained the knowledge of how to harvest and process the rice. Most younger adults and children had only consumed manomin purchased from stores served at cultural events in the region.

Restoring manomin in Wabaseemoong was not just implementing a set of technical procedures for re-establishing a habitat or population of *Zizania palustris*, but the reweaving of a set of relations with manomin (Kuzivanova, 2016; Kuzivanova and Davidson-Hunt, 2017). This required, as a first step, locating manomin fields that were not affected by dam regulation or by waters impacted by mercury. The Traditional Land Use Area agreement, signed with the government of Ontario, provided Wabaseemoong people with the political framework needed to use manomin fields outside of their reserved lands.

Along with the re-establishment of access to the fields, it also required reintroducing younger people to the practices associated with the harvest and processing of manomin. This began with a rice harvest camp in 2014 through 2017, during which young people harvested manomin, learning-by-doing with Elders in the community, who shared their knowledge (stories, values and techniques) related to the harvest, processing and cooking of manomin.

Figure 12.2 Marvin McDonald, Resources Information Officer for the Wabaseemoong Traditional Land Use Area, showing manomin grain hand harvested on the Scott River.

Source: photo V. Kuzivanova.

This approach to restoring manomin expresses a relational perspective in which restoration recentres manomin within a set of appropriate political, economic, social and cultural relations and, more generally, the life of Wabaseemoong people. Elders recognize skilled restorationists as 'manomin ogimaa', or 'wild rice boss', which, from an Anishinaabe perspective, recognizes the duty a person takes on for the care of the political, economic, social and cultural practices associated with another being with whom people co-journey on their life's pathway. This approach ensures not only that manomin has the necessary conditions to persist within the landscape (restoration in the narrow sense) but also that manomin again becomes a nutritious food supporting the social fabric of the community, is recognized within ceremony and land-based values, and becomes part of the contemporary stories of Wabaseemoong people today and into the future. Restoring manomin, in this sense, is integral to decolonizing processes linked to broader socio-ecological movements of Aboriginal resurgence, reconciliation and recognition of Treaty (rights), and healing of people, the land and relationships of place (Corntassel and Bryce, 2012). While proceeding through the steps of a restoration initiative, this experience supports the approach central to Indigenous perspectives on what is part of a coupled socio-ecological restoration process.

keystone or conservation umbrella populations) in need of specific research/monitoring and management attention. This process can also identify which species, as socio-ecological system (SES) degraders (challenges/threats/stressors), might need to be controlled or eradicated, or those that are SES integrators to be retained, promoted or reintroduced. For example, conservation efforts may emphasize linking restoration and recovery of forests to provide connectivity between protected areas, whereas indigenous cultures may focus on the importance of sacred sites connecting areas that host higher levels of biophysical and ecological diversity with culturally based conservation tenets governing the limited harvest of rare or endemic species.

Defining strategies forming the restoration framework

Forest landscape restoration programmes may use assessment of current, past and applicable 'reference' landscape states or desired conditions for informing the selection of appropriate restoration actions. In this section, we discuss the desired targets for the condition of landscape that represent the outcomes of eco-cultural revitalization efforts, management actions and stewardship practices at applicable scales. Considerations for knowledge integration and protection measures for Traditional knowledge are also offered.

FLR strategies informed by Traditional knowledge and integrating indigenous stewardship can assist with longer-term and broader-scale recovery of socio-ecological systems that support the production of ecosystem services, as well as economic and food security of indigenous and rural communities (cf. van Oosten, 2013; Parrotta *et al.*, 2015; Sterling *et al.*, 2017; Díaz *et al.*, 2018). For example, many indigenous communities (and arguably society at large) rely on forest ecosystems for provision of water and forest products. Economic development and sustainable forest management for communities are at the heart of many FLR discussions – and not just for indigenous communities.

What can be expected to be restored, and at what scale and over what anticipated time frame (years, decades or centuries), needs to be defined. Several FLR programmes have utilized results chains or modelling in the development of land use scenarios (Hawkins and Selman, 2002; Sisk *et al.*, 2006). Spatially explicit modelling and mapping of current versus desired future landscape conditions provides partners with the opportunity to evaluate potential trade-offs associated with various FLR strategies. In some cases, indigenous communities already have elaborate maps of where, when and how actions are done on the landscape. Depending on locality, it may be less a matter of developing new maps than of integrating existing maps (e.g. geographic information system (GIS) data layers), addressing values and resources of interest, with agreed-upon protocols for spatially explicit depiction (scale or specificity) across the landscape (see Lewis and Sheppard, 2006).

The development of possible restoration trajectories to achieve short-term and long-term goals (including models, time frames and maps), can assist managers and stakeholders (the public, indigenous communities) with the evaluation of different courses of action linked to conservation targets or the production of ecosystem services (Brown, 2005). Reconciliation of land use options to achieve targets based on shared values can assist with building trust in negotiating options and scenarios to accomplish FLR objectives. This process can inform the evaluation of the feasibility of specific goals while meeting shared needs or reconciling conflicting demands for particular ecosystem services (e.g. water and forest products) from the landscape. Several FLR partnerships (or collaboratives) have an identified set of goals, strategies and tactics for each zone (landscape jurisdictions, administrative types or ownerships) and contributing factors (challenges/stressors) in the landscape (Steen-Adams *et al.*, 2017). Such an approach can provide a framework for the prioritization of where, when and how FLR activities will take place on higher-value or higher-interest areas of the landscape (Uribe *et al.*, 2014).

For example, in the case of the Western Klamath Restoration Partnership in California (see Box 12.4), each component of the plan identifies required or anticipated resources, partner capacity, required knowledge/technical expertise, and time frame for allocated funds, and quantified targets are derived from the analysis. The plan includes statements of collaboratively developed commitments, with a vision statement owned by all

Box 12.4 The Western Klamath Restoration Partnership (WKRP), northern California, USA

WKRP is a collaborative land and fire management effort between tribal, federal and non-governmental stakeholders located in northwestern California. It is based on 20 years of collaborative work between diverse partners, ultimately forming the WKRP in 2013. After a century of fire exclusion and even-aged forest management, forests in the Western Klamath Mountains are in need of ecological restoration treatments to restore fire process and function. The WKRP seeks to build trust and a shared vision for restoring fire resilience at the landscape scale. The past century of governmental-enforced fire exclusion of natural and cultural burning has severely impacted water supplies, forest health, communities, cultural resources and threatened species. WKRP formulated shared values to guide principles of FLR. These values are: Fire-adapted communities; Restored fire regimes; Healthy river systems; Resilient biodiverse forests/plants and animals; Sustainable local economies; and Cultural and community vitality. Such values influence restoration strategies. WKRP development was facilitated by the Open Standards process and has brought diverse stakeholders together to accomplish work by identifying Zones of Agreement where all parties agree on upslope restoration needs. The geographical prioritization of treatments represents the overlap of multiple values and interests across the landscape. The implementation of various treatments demonstrates the partners' shared values.

WKRP created a plan for restoring fire resilience at the landscape scale, founded upon Traditional knowledge and practices, linking to concepts outlined in the United States Federal National Cohesive Wildland Fire Management Strategy (Resilient landscapes, Fire-adapted communities, Safe and effective wildfire response). These include focusing treatment around communities, transportation routes and specific geographic landscape features (e.g. strategic ridges) to manage fire to accomplish resource objectives.

This plan incorporates ecological, economic, social and cultural values spatially across a 486,000 ha landscape to determine where restoration treatments would yield the most beneficial results with the lowest impact. Restoring fire processes can create resilient landscapes, providing integral resources and services to ecosystems and communities. These include improved water quality and yield, healthy streams and aquatic populations, increased wildlife and plant diversity, eco-cultural revitalization and a sustainable supply of forest products. Local tribes and the general public depend on the landscape for a variety of social, economic and ecological factors. Treatments proposed by the WKRP aim to produce sustainable beneficial outcomes, creating forests and communities that are more resilient to stressors ranging from recent extreme wildfires and invasive species infestations to climate change and drought. With more than 90% of the nearly 0.5 million ha WKRP area in publicly owned federal national forest lands, there is an opportunity to collaboratively pursue significant large-scale ecological forest restoration. The landscape today tells a story of the traditional human/fire relationships of our past – and can guide future fire management strategies. The Karuk

Figure 12.3 Restoring resilience and functional values through fire management practices based on Traditional knowledge.

Notes

Top: (a) At Upper Rogers Creek, which includes variable-age Douglas fir/tanoak 'mixed conifer-evergreen' forest (foreground and ridge), and plantations (middle of photo), 2017 wildfires resulted in significant damage to the natural forest canopies as well as plantation areas. (Photo: F. Lake, USDA Forest Service.)

Bottom: (b) Frank Lake showing tribal students and WKRP workers a partially restored research plot of tanoak acorn and evergreen huckleberry (*Notholithocarpus densiflorus* (Hook. & Arn.) Manos, Cannon & S.H. Oh and *Vaccinium ovatum* Pursh.). This traditional food resource patch is becoming more resilient and functional for a range of cultural ecosystem services. The next restoration treatment would be an autumn understory burn to reduce additional surface fuels, control acorn insect pests, promote regrowth of huckleberry bushes, and enhance other forage plants for wildlife. (Photo: E. Knapp, USDA Forest Service.)

Tribe has refined these strategies over thousands of years to maximize diversity, resiliency and resource production. The WKRP seeks to restore these traditional practices in a modern context, enhanced by Western science, in order to restore and maintain these critical ecosystem processes, including wildland fire, to achieve multiple resource objectives. This is depicted by evaluating the condition of valued resources within planned treatment units, an emphasis on indigenous focal species, and collaboratively developed prescriptions that support the achievement of shared values with project implementation.

participants that reflects a shared perspective on what the larger FLR process, outcomes and so on mean to each member of the planning team. Alignment of the FLR vision (shared values informing Zones of Agreement) among user groups/stakeholders is important. The plan would also include a knowledge and data sharing/ownership agreement on how Traditional knowledge is incorporated, which data is sensitive or confidential (e.g. sacred sites or endangered species), and how the various forms of information will be used and made publicly available.

It is important to note that many indigenous communities may not want to share or disclose specific knowledge, beliefs or cultural practices to the general public, given often legitimate fears that this information will be appropriated or used against them, further limiting their engagement with the landscapes being restored in their traditional territories. Building trust and formalizing methods to promote effective and respectful communication among FLR parties is essential. FLR collaboratives, therefore, need to consider institutional arrangements or agreements that protect sensitive or confidential information/data as well as an understanding for ownership of data collectively acquired through planning, implementation, research/ monitoring or other interdisciplinary methods (Gamborg *et al.*, 2012).

Implementation: reviving traditional knowledge systems and practices

Forest landscape restoration programmes rely on effective planning, efficient prioritization and cohesive teams of managers trained to achieve the desired outcomes of a planning process. In this section, we identify components of effective implementation in FLR that integrates Traditional knowledge and indigenous stewardship practices. We focus on: (1) piloting successful restoration; (2) approaches to exporting or upscaling successes; and (3) implementation of restoration plans. These steps allow new partnerships (under the collaborative model) to build on smaller successes before taking on larger-scale efforts. Such a cautious approach may benefit from developing, and in some cases evolving from, smaller-scale pilot projects undertaken at specific sites identified as achievable (multiple factors lining up to support success) or because specific conditions are favourable to the realization of restoration goals (intersection of ownerships, stabilization of critical habitat, or location of priority cultural sites). While scaling up from this smaller-scale pilot phase may be viewed as essential to achieving identified conservation targets and restoration goals, these early successes, even if small scale, may represent meaningful achievements, depending on context, and serve to strengthen partner relationships.

Because of the complex history of land tenure, jurisdictions, governance, and inclusivity (participation) of stakeholders, trust and confidence built during planning stages will be tested during project implementation, and so starting at smaller geographic scales may allow rapid corrections and adaptation to the evolution of partnership thinking in response to visualizing on-the-ground demonstrations (van Oosten, 2013). For indigenous communities, going back to traditionally (historically, before colonization or indigenous relocation) managed sites may be more efficient in some cases than developing new pilot sites. These localities, despite the complexity of their socio-political and ecological history, may be most effective for demonstrating the legacy of former traditional management systems and how this can inform current and future strategies across the landscape. An important part of implementation addresses relationship-building, not only among those involved but also between people and place (Kimmerer, 2011). The foundation of cooperation with pilot projects demonstrates capacity that may facilitate cooperative action at larger, more complex scales of the landscape. Implementation of larger-scale actions often requires more socio-political negotiations, policy and governance authority and support, recruitment and evaluation of contributing knowledge systems, funding and personnel capacity, with opportunities for sharing resources across jurisdictions (van Oosten, 2013). Lessons learned from initial pilot project results, both successes and failures, inform adaptive management in FLR implementation (Berkes *et al.*, 2000; O'Connor *et al.*, 2005).

Evaluation, corrective actions and adaptive management moving towards fully restored coupled human and natural systems

The emphasis in this section is on the role of Traditional knowledge approaches to monitoring, criteria and indicators, and evaluation of achieving sustainable management applicable to FLR programmes. In this section, we identify and focus on monitoring changes in resource condition and the production of desired ecosystem goods and services. We offer a conceptual model (Figure 12.4) linking ecological and socio-ecological processes across scales to evaluate their influences on FLR strategies.

Successful scaling up from pilot- to landscape-scale projects requires evaluation of various meaningful metrics or indicators that reflect the shared values (social, economic, ecological) of restoration partners and,

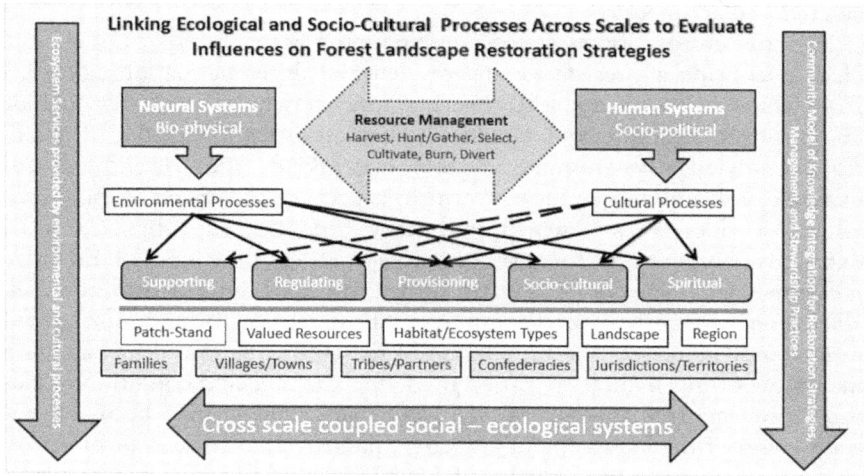

Figure 12.4 The Processes and Scale figure depicts a way to frame the interactions of Natural (bio-physical) and Human (socio-political) influences and associated processes for coupled socio-ecological systems. There are environmental and cultural influences on the production of ecosystem services. Some of these processes have greater and lesser degrees of influence and direct or indirect effects on ecosystem services, as indicated by the thickness of solid and dashed lines. The bottom boxes above the Cross scale arrows depict (left to right) local area (Patch-Stand/Families) to broader regional extent (Landscapes/Jurisdictions or Territories) of natural and human system interactions of resource management needed to support FLR strategies.

more broadly, of all stakeholders (i.e. landscape-scale ecosystem service/ goods beneficiaries). The restoration trajectory reappraisal, as informed by Traditional and Western knowledge systems' criteria, should be evaluated according to previously agreed-upon qualitative and quantitative benchmarks or metrics reflecting advancement towards or achievement of broader values. At this stage, the role of Traditional knowledge can inform approaches to monitoring changes in resource condition based on socio-culturally meaningful criteria for evaluating indicators that represent shared values (Sterling *et al.*, 2017). By incorporating indigenous or local community measures of success or failure, partners can collaboratively develop a monitoring plan and incorporate research to objectively evaluate treatments. The incorporation of monitoring activities before, during and subsequent to restoration treatments/practices at multiple scales is critical to the implementation of corrective actions when and where needed, or may provide validation check points for the FLR process. Ultimately, while there are some influences on the restoration of forested landscapes that management actions can address and have the ability to influence, other stressors or agents of change (e.g. climate) will have to be identified, considered and adapted to.

Effective restoration strategies anticipate the potential effects of climate change, considering resistance and resilience at different spatial scales (i.e. from stand to landscape). Recovering forests, whose attributes include composition, structure and functional characteristics, coupled with traditional knowledge systems and practices, may take decades to centuries. For example, cultural dependency on critical aspects of the resources only provided by older forests, which may have been degraded by large-scale extractive commercial forestry practices, may require special adaptive measures that incorporate climate mitigation and cultural resource considerations. In such cases, an appropriate objective (or monitoring criterion) may be to retain a proportion of the landscape in old forests, with indicators of this habitat condition involving size, age and condition of cultural keystone tree species. Another value or criterion may be to have a proportion of the landscape in an active, intensively managed agroforestry gradient for forests or agricultural lands. Various restoration treatments can, at different scales and locations, facilitate the advancement of ecological processes and socio-cultural practices that promote such desired conditions. Some of these approaches and desires may be mutually supportive, while others may conflict at a given scale or for particular habitats of the landscape. A successful FLR programme that evaluates outcomes of processes and actions at various stages can support the recoupling of human and natural systems through the promotion of Traditional and Western knowledge-based stewardship practices that enhance provision of ecosystem services from the landscape of interest.

Given a range of future uncertainties, partners will need to consider the production of desired ecosystem goods and services that support economic as well as cultural non-market forest products. The process of monitoring and evaluation of what was planned versus the actual outcomes of restoration treatments and strategies will require finding cohesion in the shared values, interests and management strategies of diverse user groups, local and indigenous, as well as the broader public and societies for landscapes (Lamb *et al.*, 2012; Sterling *et al.*, 2017). A successful FLR programme evaluates the attainment of processes and actions, at various stages, that support the recoupling of human and natural systems through promotion of Traditional and Western knowledge-based stewardship practices to enhance the provision of ecosystem services from the landscape of interest.

Conclusions

In this chapter, we explored the process of integrating multiple knowledge systems in FLR. FLR strategies that integrate Western and Traditional knowledge systems can improve our understanding of coupled social-ecological systems. A collective understanding of 'what degradation means and is' is important. Well-integrated approaches to FLR can be effective at reducing or mitigating drivers of landscape degradation that affect both the socio-cultural and ecological communities. Sustainable harvesting and management practices emphasize the recovery of degraded forests, applicable cultural practices, and ecological processes to enhance the production of ecosystem goods and services. With such integration, local and indigenous community values associated with habitats and resources are balanced with the broader interests of external stakeholders/the public and society (Brown, 2005). The various steps of the FLR planning process discussed in this chapter provide a means to address the cultural, social and economic facets of sustainable management of landscapes and the integration of knowledge systems. There are important challenges but also opportunities associated with, and required for, knowledge system integration. We suggest that contemporary professionals leading efforts to integrate Traditional knowledge into FLR should acknowledge, examine and seek to better understand colonial and other historical legacies within current restoration approaches (Trosper *et al.*, 2012b).

Understanding which type of partnership model can be used, and which FLR strategies are feasible, in a given landscape allows more effective integration of Traditional knowledge into the visioning, planning, overall decision-making and implementation. A mutual learning process among participants can identify shared values and interests, and establish a common language and perspective on land, resources, and management strategies, building the trust necessary for successful outcomes. Scaling up from smaller-scale pilot projects may be essential to achieving identified conservation targets and restoration goals. Insights gained from the case studies presented and the approaches discussed in this chapter highlighted efforts to accommodate Traditional knowledge to a greater extent than previously considered. Various restoration treatments can, at different scales and locations, facilitate the advancement of ecological processes and socio-cultural practices that promote desired conditions to achieve FLR.

Success in collaborative management for FLR ultimately requires: (i) integrating management of the biophysical landscape with development needs of local residents; (ii) embracing the complexity and challenges associated with pluralistic decision-making and (iii) finding ways to legitimize and facilitate a transition from top-down decision-making to bottom-up collaborative approaches. Partnerships can utilize Traditional and Western knowledge for problem conceptualization, planning, strategy development, implementation, management, research/monitoring and adaptation, which collectively can result in long-term collaborative restoration of landscapes and their ongoing stewardship, ensuring that they contribute benefits to local and indigenous peoples and the wider societies within which they persist (Sterling *et al.*, 2017; Díaz *et al.*, 2018).

Acknowledgements

The authors thank Marvin McDonald and Valeria Kuzivanova for their contributions to the case study presented in Box 12.3.

References

Andrade, G. S. M. and Rhodes, J. R. (2012), 'Protected areas and local communities: an inevitable partnership toward successful conservation strategies?' *Ecology and Society*, vol 17, no 4, p. 14.

Berkes, F. (2007), 'Community-based conservation in a globalized world', *Proceedings of the National Academy of Sciences*, vol 104, pp. 15188–15193.

Berkes, F. (2009), 'Evolution of co-management: role of knowledge generation, bridging organizations and social learning', *Journal of Environmental Management*, vol 90, pp. 1692–1702.

Berkes, F., Colding, J. and Folke, C. (2000), 'Rediscovery of traditional ecological knowledge as adaptive management', *Ecological Applications*, vol 10, pp. 1251–1262.

Berkes, F. and Davidson-Hunt, I. J. (2006), 'Biodiversity, traditional management systems, and cultural landscapes: examples from the boreal forest of Canada', *International Social Science Journal*, vol 187, pp. 35–47.

Berkes, F., Kofinas, G. P. and Chapin III, F. S. (2009), 'Conservation, community, and livelihoods: sustaining, renewing, and adapting cultural connections to the land', in Chapin III, F. S., Kofinas, G. P. and Folke, C., eds, *Principles of ecosystem stewardship* (pp. 129–147), Springer, New York.

Boedhihartono, A. K. and Sayer, J. (2012), 'Forest landscape restoration: restoring what and for whom?', in Stanturf, J., Lamb, D. and Madsen, P. eds, *Forest landscape restoration*, vol 15 (pp. 309–323), Springer, Dordrecht.

Botkin, D. B. (2004), *Our natural history: The lessons of Lewis and Clark*, Oxford University Press on Demand.

Bowman, D. M., Balch, J., Artaxo, P., Bond, W. J., Cochrane, M. A., D'antonio, C. M., DeFries, R., Johnston, F. H., Keeley, J. E., Krawchuk, M. A. and Kull, C. A. (2011), 'The human dimension of fire regimes on Earth', *Journal of Biogeography*, vol 38, no 12, pp. 2223–2236.

Brown, K. (2005), 'Addressing trade-offs in forest landscape restoration', in Mansourian, S., Dudley, N. and Vallauri, D., eds, *Forest restoration in landscapes* (pp. 59–64), Springer, New York.

Charnley, S., Fischer, A. P. and Jones, E. T. (2007), 'Integrating traditional and local ecological knowledge into forest biodiversity conservation in the Pacific Northwest', *Forest Ecology and Management*, vol 246, no 1, pp. 14–28.

Chase, A. (1986), *Playing God in Yellowstone. The destruction of America's first national park*, Harcourt Brace and Co., New York.

Chazdon, R. L., Brancalion, P. H., Lamb, D., Laestadius, L., Calmon, M. and Kumar, C. (2017), 'A policy-driven knowledge agenda for global forest and landscape restoration', *Conservation Letters*, vol 10, no 1, pp. 125–132.

Corntassel, J. and Bryce, C. (2012), 'Practicing sustainable self-determination: Indigenous approaches to cultural restoration and revitalization', *Brown Journal of World Affairs*, vol 18, no 2, pp. 151–162.

Cronon, W. (1996), 'The trouble with wilderness: or, getting back to the wrong nature', *Environmental History*, vol 1, no 1, pp. 7–28.

Davidson-Hunt, I. J. and Berkes, F. (2003), 'Learning as you journey: Anishinaabe perception of social-ecological environments and adaptive learning', *Conservation Ecology*, vol 8, no 1, p. 5, available at: www.ecologyandsociety.org/vol8/iss1/art5/.

Davidson-Hunt, I. J., Jack, P., Mandamin, E. and Wapioke, B. (2005), 'Iskatewizaagegan (Shoal Lake) plant knowledge: an Anishinaabe (Ojibway) ethnobotany of northwestern Ontario', *Journal of Ethnobiology*, vol 25, no 2, pp. 189–227.

Davidson-Hunt, I. J. and O'Flaherty, R. M. (2007), 'Researchers, indigenous peoples and place-based learning communities', *Society and Natural Resources*, vol 20, no 4, pp. 1–15.

Díaz, S., Pascual, U., Stenseke, M., Martín-López, B., Watson, R. T., Molnár, Z., Hill, R., Chan, K. M., Baste, I. A., Brauman, K. A. and Polasky, S. (2018), 'Assessing nature's contributions to people', *Science*, vol 359, no 6373, pp. 270–272.

Donatuto, J., Grossman, E. E., Konovsky, J., Grossman, S. and Campbell, L. W. (2014), 'Indigenous community health and climate change: integrating biophysical and social science indicators', *Coastal Management*, vol 42, no 4, pp. 355–373. doi: 10.1080/08920753.2014.923140.

Dudley, N., Mansourian, S. and Vallauri, D. (2005), 'Forest landscape restoration in context', in Mansourian, S., Vallauri, D. and Dudley, N., eds, *Forest restoration in landscapes: beyond planting trees* (pp. 3–7), Springer, New York.

Gamborg, C., Parsons, R., Puri, R. K. and Sandøe, P. (2012), 'Ethics and research methodologies for the study of traditional forest-related knowledge', in Parrotta, J. and Trosper, R. L., eds, *Traditional forest-related knowledge: sustaining communities, ecosystems and biocultural diversity*, World Forest Series vol. 12 (pp. 535–562), Dordrecht, Springer.

Garibaldi, A. and Turner, N. (2004), 'Cultural keystone species: implications for ecological conservation and restoration', *Ecology and Society*, vol 9, no 3, p. 1.

Giardina, C., Litton, C. M., Thaxton, J. M., Cordell, S., Hadway, L. and Sandquist, D. R. (2007), 'Science driven restoration: a candle in a demon haunted world – response to Cabin (2007)', *Restoration Ecology*, vol 15, pp. 171–176.

Gilmour, D. (2016), *Forty years of community-based forestry: a review of its extent and effectiveness*, Food and Agriculture Organization of the United Nations Forestry Paper 176, FAO, Rome.

Grantham, H. S., Bode, M., McDonald-Madden, E., Game, E. T., Knight, A. T. and Possingham, H. P. (2010), 'Effective conservation planning requires learning and adaptation', *Frontiers in Ecology and the Environment*, vol 8, no 8, pp. 431–437.

Grenier, L. (1998), *Working with indigenous knowledge – a guide for researchers*, International Development Research Centre, Ottawa, Canada.

Hall, T. D. and Fenelon, J. V. (2016), *Indigenous peoples and globalization*, Routledge, New York.

Hawkins, V. and Selman, P. (2002), 'Landscape scale planning: exploring alternative land use scenarios', *Landscape and urban planning*, vol 60, no 4, pp. 211–224.

Hessburg, P. F., Churchill, D. J., Larson, A. J., Haugo, R. D., Miller, C., Spies, T. A., Malcolm P. North, M. P., Povak, N. A., Belote, R. T., Singleton, P. H., Gaines, W. L., Keane, R. E., Aplet, G. H., Stephens, S. L., Morgan, P., Bisson, P. A., Rieman, B. E., Salter, R. B. and Reeves, G. H. (2015), 'Restoring fire-prone

Inland Pacific landscapes: seven core principles', *Landscape Ecology*, vol 30, no 10, pp. 1805–1835.

Higgs, E. S. (2005), 'The two-culture problem: ecological restoration and the integration of knowledge', *Restoration Ecology*, vol 13, pp. 159–164.

Joseph, G. and Mansourian, S. (2005), 'Restoring landscapes for traditional cultural values', in *Forest restoration in landscapes* (pp. 233–238), Springer, New York.

Kealiikanakaoleohaililani, K. and Giardina, C. P. (2016), 'Embracing the sacred: an indigenous framework for tomorrow's sustainability science', *Sustainability Science*, vol 11, no 1, pp. 57–67. doi: 10.1007/s11625-015-0343-3.

Kimmerer, R. (2011), 'Restoration and reciprocity: the contributions of traditional ecological knowledge', in Egan, D., Hjerpe, E. E. and Abrams, J., eds, *Human dimensions of ecological restoration: integrating science, nature, and culture* (pp. 257–276), Island Press, Washington, DC.

Kimmerer, R. W. and Lake, F. K. (2001), 'The role of indigenous burning in land management', *Journal of Forestry*, vol 99, no 11, pp. 36–41.

Klooster, D. J. (2002), 'Toward adaptive community forest management: integrating local forest knowledge with scientific forestry', *Economic Geography*, vol 78, no 1, pp. 43–70.

Kuzivanova, V. (2016), *Restoring manomin (wild rice): a case study with Wabaseemoong Independent Nations, Ontario*. M.N.R.M. Thesis, The University of Manitoba, Canada.

Kuzivanova, V. and Davidson-Hunt, I. J. (2017), 'Biocultural design: harvesting manomin with Wabaseemoong Independent Nations', *Ethnobiology Letters*, vol 8, no 1, pp. 23–30.

Lake, F. K., Wright, V., Morgan, P., McFadzen, M., McWethy, D. and Stevens-Rumann, C. (2017), 'Returning fire to the land: celebrating traditional knowledge and fire', *Journal of Forestry*, vol 115, no 5, pp. 343–353.

Lamb, D., Stanturf, J. and Madsen, P. (2012), 'What is forest landscape restoration?' in *Forest landscape restoration* (pp. 3–23), Springer, Dordrecht.

Lewis, J. L. and Sheppard, S. R. (2006), 'Culture and communication: can landscape visualization improve forest management consultation with indigenous communities?', *Landscape and Urban Planning*, vol 77, no 3, pp. 291–313.

Long, J. W., Anderson, M. K., Quinn-Davidson, L., Goode, R. W., Lake, F. K. and Skinner, C. N. (2016), 'Restoring California black oak ecosystems to promote tribal values and wildlife', *Gen. Tech. Rep. PSW GTR-252*. Albany, CA, US Department of Agriculture, Forest Service, Pacific Southwest Research Station. 110.

Mann, C. C. (2002), '1491', *The Atlantic Monthly*, March 2002. 13 p.

Martinez, D. (2014), 'Traditional ecological knowledge, traditional resource management and silviculture in ecocultural restoration of temperate forests', in Bozzano, M., Jalonen, R., Thomas, E., Boshier, D., Gallo, L., Cavers, S., Bordacs, S., Smith, P. and Loo, J., eds, *Genetic considerations in ecosystem restoration using native tree species: state of the world's forest genetic resources* (pp. 109–120), Food and Agriculture Organization of the United Nations and Biodiversity International, Rome.

Nash, R. (2014), *Wilderness and the American mind*, Yale University Press, London.

O'Connor, S., Salafsky, N. and Salzer, D. (2005), 'Monitoring forest restoration projects in the context of an adaptive management cycle', in *Forest restoration in landscapes* (pp. 145–149), Springer, New York.

Parrotta, J. A., Dey de Pryck, J., Darko Obiri, B., Padoch, C., Powell, B. and Sandbrook, C. (2015), 'The historical, environmental and socio-economic context of forests and tree-based systems for food security and nutrition', Chapter 3 (pp. 71–134) in Vira, B., Mansourian,S. and Wildburger, C., eds, *Forests and food: addressing hunger and nutrition across sustainable landscapes*, Open Book Publishers, Cambridge, UK. http://dx.doi.org/10.11647/OBP.0085.

Parrotta, J. A. and Trosper, R. L., eds. (2012), *Traditional forest-related knowledge: sustaining communities, ecosystems and biocultural diversity*. World Forest Series vol. 12, Springer, Dordrecht.

Pillsbury, R. W. and McGuire, M. A. (2009), 'Factors affecting the distribution of wild rice (*Zizania palustris*) and the associated macrophyte community', *Wetlands*, vol 29, no 2, pp. 724–734.

Ribot, J. C., Agrawal, A. and Larson, A. M. (2006), 'Recentralizing while decentralizing: how national governments reappropriate forest resources', *World Development*, vol 34, pp. 1864–1886.

Sisk, T. D., Prather, J. W., Hampton, H. M., Aumack, E. N., Xu. Y. and Dickson, B. G. (2006), 'Participatory landscape analysis to guide restoration of ponderosa pine ecosystems in the American Southwest', *Landscape and Urban Planning*, vol 78, no 4, pp. 300–310.

Smith, L. T. (2012), *Deconstructing methodologies: research and indigenous peoples*, Zed Books, London.

Steen-Adams, M., Charnley, S. and Adams, M. (2017), 'Historical perspective on the influence of wildfire policy, law, and informal institutions on management and forest resilience in a multiownership, frequent-fire, coupled human and natural system in Oregon, USA', *Ecology and Society*, vol 22, no 3, article 23. https://doi.org/10.5751/ES-09399-220323.

Steen-Adams, M. M., Langston, N., Adams, M. D. and Mladenoff, D. J. (2015), 'Historical framework to explain long-term coupled human and natural system feedbacks: application to a multiple-ownership forest landscape in the northern Great Lakes region, USA', *Ecology and Society*, vol 20, no 1, article 28. http://dx.doi.org/10.5751/ES-06930-200128.

Steen-Adams, M. M., Mladenoff, D. J., Langston, N. E., Liu, F. and Zhu, J. (2011), 'Influence of biophysical factors and differences in Ojibwe reservation versus Euro-American social histories on forest landscape change in northern Wisconsin, USA' *Landscape Ecology*, vol 26, no 8, pp. 1165–1178.

Sterling, E. J., Filardi, C., Toomey, A., Sigouin, A., Betley, E., Gazit, N., Newell, J., Albert, S., Alvira, D., Bergamini, N. and Blair, M. (2017), 'Biocultural approaches to well-being and sustainability indicators across scales', *Nature Ecology & Evolution*, vol 1, no 12, pp. 1798–1806.

Stevenson, M. G. and Webb, J. (2003), 'Just another stakeholder? First Nations and sustainable forest management in Canada's boreal forest', in Burton, P. J., Messier, C., Smith, D. W. and Adamowicz, W. L., eds, *Towards sustainable management of the boreal forest* (pp. 65–112), NRC Research Press, Ottawa, Ontario, Canada.

Stewart, O. (2002), *Forgotten fires: Native Americans and the transient wilderness*, University of Oklahoma Press, Norman, OK.

Trosper, R. L., Clark, F., Gerez-Fernandez, P., Lake, F., McGregor, D., Peters, C. M., Purata, S., Ryan, T., Thomson, A., Watson, A. and Wyatt, S. (2012a), 'North America', in Parrotta, J. A. and Trosper, R. L., eds, *Traditional forest-related knowledge: sustaining communities, ecosystems and biocultural diversity* (pp. 157–201), Springer, Dordrecht.

Trosper, R. L., Parrotta, J. A., Agnoletti, M., Bocharnikov, V., Feary, S. A., Gabay, M., Gamborg, C., García Latorre, J., Johann, E., Laletin, A., Lim, H. F., Oteng-Yeboah, A., Pinedo-Vasquez, M. A., Ramakrishnan, P. S. and Youn, Y. C. (2012b), 'The unique character of traditional forest-related knowledge: threats and challenges ahead', inParrotta, J. A. and Trosper, R. L., eds, *Traditional forest-related knowledge: sustaining communities, ecosystems and biocultural diversity* (pp. 563–588), Springer, Dordrecht.

UNDRIP (2007), *United Nations Declaration on the Rights of Indigenous Peoples*, www.un.org/development/desa/indigenouspeoples/declaration-on-the-rights-of-indigenous-peoples.html (Accessed 13 February 2018).

Uprety, Y., Asselin, H., Bergeron, Y., Doyon, F. and Boucher, J.-F. (2012), 'Contribution of traditional knowledge to ecological restoration: practices and applications', *Ecoscience*, vol 19, no 3, pp. 225–237. doi:10.2980/19-3-3530.

Uprety, Y., Asselin, H. and Bergeron, Y. (2013), 'Cultural importance of white pine (*Pinus strobus* L.) to the Kitcisakik Algonquin community of western Quebec, Canada', *Canadian Journal of Forest Research*, vol 43, pp. 544–551.

Uprety, Y., Asselin, H., Bergeron, Y. and Mazerolle, M. J. (2014), 'White pine (*Pinus strobus* L.) regeneration dynamics at the species' northern limit of continuous distribution', *New Forests*, vol 45, pp. 131–147.

Uprety, Y., Asselin, H. and Bergeron, Y. (2017), 'Preserving ecosystem services on indigenous territory through restoration and management of a cultural keystone species', *Forests*, vol 8, no 6, p. 194.

Uribe, D., Geneletti, D., del Castillo, R. F. and Orsi, F. (2014), 'Integrating stakeholder preferences and GIS-based multicriteria analysis to identify forest landscape restoration priorities', *Sustainability*, vol 6, no 2, pp. 935–951.

Vallauri, D., Aronson, J. and Dudley, N. (2005), 'An attempt to develop a framework for restoration planning', in Mansourian, S., Vallauri, D. and Dudley, N., eds, *Forest restoration in landscapes* (pp. 65–70), Springer, New York.

van Oosten, C. (2013), 'Restoring landscapes – governing place: a learning approach to forest landscape restoration', *Journal of Sustainable Forestry*, vol 32, no 7, pp. 659–676. doi: 10.1080/10549811.2013.818551.

Vennum, T. Jr. (1988), *Wild rice and the Ojibway people*, Minnesota Historical Society, St. Paul, MN.

von der Porten, S., de Loë, R. C. and McGregor, D. (2016), 'Incorporating Indigenous knowledge systems into collaborative governance for water: challenges and opportunities', *Journal of Canadian Studies*, vol 50, no 1, pp. 214–243.

von der Porten, S., de Loë, R. C. and Plummer, R. (2015), 'Collaborative environmental governance and indigenous peoples: recommendations for practice', *Journal of Environmental Practice*, vol 17, no 2, pp. 134–144.

von der Porten, S. and de Loë, R. C. (2014), 'How collaborative approaches to environmental problem solving view indigenous peoples: a systematic review', *Society & Natural Resources*, vol 27, no 10, pp. 1040–1056.

Wallman, D., Wells, E. C. and Rivera-Collazo, I. C. (2018), 'The environmental legacies of colonialism in the northern Neotropics: introduction to the special issue', *Environmental Archaeology*, vol 23, no. 1, pp. 1–3.

Williams, R. A. (1990), *The American Indian in Western legal thought: the discourses of conquest*, Oxford University Press, Oxford.

Part IV
Synthesis and conclusions

13 Putting the pieces together

Integration for forest landscape restoration implementation

John Parrotta and Stephanie Mansourian

Through this volume, we sought to expand our reach beyond the traditional forest landscape restoration (FLR) community and related disciplines of forestry and ecology. As we outlined the chapters for the book, we purposefully stretched the boundaries of current research on FLR. Chapters 1 and 2 have argued our case for an integrated approach, explicating, on the one hand, some of the factors that have led us to prioritize forest restoration around the globe. And, on the other hand, we critically examined some of the efforts to date labelled as 'restoration' in the broadest sense. In seeking a way forward, our starting point was that FLR necessarily involves people and that not much has been written about people in FLR (notable exceptions being Aronson *et al.*, 2010; Egan *et al.*, 2011; Emborg *et al.*, 2012). In recognition of this, we decided to dedicate a chapter specifically to stakeholders (Chapter 9), which explores the diversity of actors in FLR. Understandings of what constitutes forest landscape degradation, as well as knowledge about the landscape and forests, that are currently applied in restoration interventions are generally limited to Western science, yet there exists a wealth of relevant knowledge among indigenous and local communities around the world that can be integrated into FLR planning, implementation and monitoring. So, we also explored the distinctive features and potential value of traditional knowledge in relation to forest and landscape management (Chapter 3), as well as opportunities to integrate diverse knowledge systems in FLR initiatives (Chapter 12). Because people take decisions in the landscape, for better or worse, that reflect formal or informal governance arrangements, we dedicated a chapter to polycentric governance that recognizes the role of diverse actors in decision-making (Chapter 11). Yet, inequalities have characterized landscape interventions, and we identify some pitfalls in this respect from other land use interventions (Chapter 4).

We acknowledged that the *scale* of FLR raises specific challenges. The landscape approach and landscape-scale interventions are increasingly being advocated in several policymaking circles. There is a close link between these and FLR, in terms of both spatial scale and human/ecological trade-offs; a relationship we explore in Chapter 6. One significant challenge brought about by this larger scale is that of land and forest tenure, as more stakeholders means that more tenure and rights claims are likely (Chapter 10). The influences on the landscape from political, socio-economic and ecological processes operating at both smaller and larger scales (both within and beyond landscapes) were also of interest to us.

Chapter 5 explores linkages between the broader concept of social-ecological systems and FLR. Looking beyond restoration at opportunities to accommodate different land uses in the landscape and outside it, we explored the land sparing/land sharing debate in Chapter 7. And, repositioning people at the centre of FLR implementation, we considered the specific case of agroecological approaches in post-conflict landscapes in Colombia (Chapter 8) as a means of respecting local people's traditions and supporting them, while managing land and forest in such a way that they can be restored to provide benefits both to local people and for biodiversity.

There have been attempts at tackling FLR and large-scale forest restoration in a relatively linear fashion (e.g. Vallauri *et al.*, 2005), even if feedback loops and adaptive management may be considered (e.g. Keenleyside *et al.*, 2012). Current efforts in international organizations are centred on designing principles for FLR (e.g. Newton *et al.*, 2012; Brancalion and Chazdon, 2017; Chazdon and Guariguata, 2018). There is also talk of standards (McGuire, 2014) and safeguards in FLR implementation (Pistorius and Kiff, 2017). For our purposes and to reach our objective – which was to expand the breadth and reach of restorationists so that they may tackle more effectively the challenge of restoring the world's deforested and degraded landscapes – we placed FLR at the centre of a series of concentric circles, and sought to learn from these larger circles to see what could help us in future efforts on FLR.

So what have we learned?

Looking back at the wealth and diversity of information emerging from the chapters in this volume, we return to the three questions we posed in the Introduction to identify some of the salient lessons emerging from this combined body of knowledge.

Question 1: what are some of the integration challenges for FLR?

In answering our first question, we identify five significant challenges.

1 Narrow silo-based approaches cannot address the diversity of issues present in landscapes being restored

Narrow approaches have created externalities, with, for example, an emphasis on carbon sequestration in reforestation schemes, ignoring the many additional benefits – such as food provision, soil stabilization and biodiversity conservation – that a more comprehensive approach to restoration or reforestation could achieve. Examples have been presented in this volume of reforestation/restoration efforts targeting only a few benefits that have failed because they did not consider local needs. Silo approaches have pitted government ministries against each other, with agriculture ministries contradicting environment ministries' priorities, infrastructure and tourism development challenging environmental priorities, among others. For example, Chapter 2 illustrated the case of Indonesia's peatlands, where in the wake of the fires of 2015, the Ministry of Public Works and Housing continued to plan for peatland drainage, while the Ministry of Environment and Forestry had restricted such practices, and the Ministry of Public Works planned to close some of the major drainage canals, even though smaller canals fall under the jurisdiction of provincial and local administrations.

Because of the long-term nature of restoration, and the changes in land use that it implies, more comprehensive strategies are required that address multiple objectives. FLR necessarily straddles sectors and therefore provides a natural opportunity for integration. Profound institutional reform is necessary that can bring together decision-making processes affecting land use so that restoration may contribute to a broader set of benefits in the context of a comprehensive and long-term plan or vision for the landscape. Yet, the complexity inherent in seeking multiple objectives is typically not addressed in attempts to restore forest landscapes that are dominated by a single sector, discipline or interest group.

2 Progress is hampered by a lack of common understanding among stakeholders concerning causes of forest loss and degradation, objectives for FLR and implementation actions

Stakeholders often disagree about both the reality and the implications of forest landscape degradation. Further, stakeholders may have divergent interpretations of the causes of forest and land degradation, in terms of either the proximal (e.g. land use practices) or underlying causes (e.g. social and economic drivers, governance failures, etc.). Such divergence is related to differences in perception, history, value systems and notions of 'development' (e.g. does conversion of forests to croplands represent broadly beneficial land use change or degradation?). These divergent views are shaped by the differential economic, social, cultural and ecological impacts that forest landscape degradation – or its reversal via FLR – has on different stakeholder groups (i.e. among 'winners' and 'losers') at different temporal and spatial scales. For example, in Chapter 9, the case of Canada's Cape Breton Highlands National Park highlighted the divergent interpretations of the landscape scale between the Mi'qmak community and Parks Canada, which has implications for the subsequent scale and choice of restoration interventions.

Not only is it difficult to reconcile different understandings in FLR implementation, but there are also divergences among priorities voiced by different stakeholders. Power inequalities may mean that those with a stronger political voice may promote their agendas at the expense of the less powerful. Indeed, FLR may create or exacerbate inequalities between groups, notably for the most vulnerable and disenfranchised, including indigenous groups and women. Given the current scale of interest in FLR, risks of exacerbating these inequalities are very real and require us to pay particular attention to them. Motivations differ among stakeholders, such that they may approach the process from diverse angles, resulting in many interpretations of the FLR purpose and process. Diverse interpretations of the problem, its impacts and the proposed solutions lead to implementation challenges and even paralysis.

3 Tenure and property rights impact on FLR in complex ways

Ownership and rights to land determine to a large extent the relationship, engagement and investment in the land. Yet, the discourse over ownership and rights is more nuanced. Insecure tenure may mean that customary rights may be flouted by more powerful interests, leading to lower investments in restoration efforts by local communities. For example, the case of the Compensatory Afforestation Fund Management and Planning Authority Act (CAMPA) in India, discussed in Chapter 4, shows how the implementation of an afforestation programme affected the rights of local people as well as tenurial arrangements through appropriation of common lands for plantation. Consequently, it has been argued that an equitable FLR process should consider such power imbalances among diverse stakeholders in FLR.

On the other hand, property rights do not necessarily need to be legally established for restoration to take place, as has been evidenced in some case studies. While land tenure and property rights are of paramount importance in landscape-scale interventions such as FLR, the argument that legal tenure is a necessary precondition may lead to limited opportunities for FLR. Instead, there is a continuum of measures between the two extremes of legality and illegality of rights to ensure that different property rights are secured. Recognizing this continuum can facilitate the definition and application of institutions, rules and policies that may provide the necessary supportive environment for FLR implementation. In Chapter 10, the example of Changting County in western Fujian Province showed that tenure reforms giving individuals or legal entities wide-reaching land use rights, combined with incentives under the Conversion of Cropland to Forest Programme (CCFP), led to an increase in forest cover by 20% over a 30-year period and also an increase in average annual farm income from US$60 to US$1110.

4 Tensions between planning the landscape-level process and implementing local actions need to be acknowledged and addressed

Ambiguity exists when seeking to clearly define objectives for a landscape, such as hectares to be restored or amounts of carbon to be sequestered, given the dynamic nature of landscapes. Short-term imperatives (political, economic) and long-term processes (ecological, social) may clash when seeking to define FLR objectives for the landscape. Designing an FLR plan may, therefore, require more built-in flexibility related to timing and feedback loops from the different nested systems comprising the landscape (and beyond). Policies that may promote some aspects of restoration in the landscape may stifle others. In such cases, more innovative means to reconcile different objectives – particularly between those of local-level and more distant stakeholders – need to be found. In Chapter 8, conflicting priorities between the state and local peasant organizations in Colombia illustrated this need for reconciliation. Whereas the latter wanted to promote agroecological practices based on traditional knowledge and associated social institutions to strengthen local productive capacities and stimulate technological innovation, the state designed policies and regulations with a view to fostering agricultural production, primarily for global commodity markets.

5 There is a risk of recentralization of forest governance

Lessons emerging from other experiences, such as previous attempts at large-scale reforestation or REDD+ (reducing emissions from deforestation and forest degradation, and the role of conservation, sustainable management of forests, and enhancement of forest carbon stocks in developing countries) interventions, have raised concerns that such approaches could lead to more centralized forest management. Requirements under diverse schemes, and particularly the association between forest restoration and carbon capture (and permanence of the sequestered carbon), have led some to question the ability of decentralized institutions, particularly informal and/or traditional ones, to achieve the desired results. The implicit risk of recentralization could erase decades of progress in community empowerment and polycentric governance. Without proper safeguards, FLR can become an excuse to wrest power from stakeholders present in the landscape and those most dependent on it. For example, the 2016 CAMPA in India was illustrated (in Chapter 4) as an example of the state forest department defining the species and choosing the land on which to replant, at the expense of local practices and wishes.

Question 2: what can we learn from other large-scale land use initiatives, frameworks or approaches?

Our second question was focused specifically on learning from existing large-scale and integrative processes. In responding to this question, we identified the following:

1 From the social-ecological system (SES) framework, we learned that...

Understanding of the roles of multiple stakeholders is essential. Landscape stakeholders are nested within scales of social and ecological interactions, which influence their power to act and their decisions. Factors across these scales have an impact at both lower and higher levels, with, for example, household decisions influencing the community and international decisions impacting households. Recognizing this interaction and interdependence helps to better consider less obvious agents in the process of FLR. This equally applies to governance, where nested systems can be identified that influence the FLR process, from household and community-level decision-making to national governance systems and international influences. Chapter 5 highlighted how trade-offs and gains will be determined by the scale of implementation of specific restoration actions.

A balance must be sought between the resilience of human and ecological systems. Different restoration measures will influence both social and ecological resilience in different ways. Resilience is understood as a fundamental feature of ecosystems, particularly in the face of climate change, but is also relevant to social systems. Considering that the landscape is integrated within nested systems, impacts on the resilience of any of these systems may have an effect on related systems. On the one hand, restoration may shift the intrinsic resilience of an SES, while on the other hand, restoration in the context of landscapes should aim for stability of the social and ecological systems. This emphasis on social-ecological resilience, and the value of SES frameworks for understanding how FLR planning and management interventions can affect resilience, is particularly relevant to the integration of knowledge systems and working with indigenous and local communities (as discussed in Chapter 12).

Embracing complexity shines a light on trade-offs. The SES framework can help to better understand the complexity implicit in FLR and bring the elements of the social and ecological systems together under one umbrella. It provides the means to dissect the complexity inherent in SES and combine disciplines to understand relationships among different variables in these systems. By doing so, SES frameworks allow us to identify and thus tackle trade-offs, supporting FLR both as a management approach and as a research tool. Such frameworks also help to reach beyond the landscape and to make connections across sectors.

2 From the landscape approach we learned that...

We must look beyond the landscape. FLR takes place within a dynamic context of political, economic, social, cultural and ecological processes operating at various spatial and temporal scales both within and beyond the landscape in question. Influences from outside the landscape include, for example, local-level institutions and larger international land use policies. Challenges have been identified related to the landscape approach in particular, because the landscape does not fit any specific jurisdictional boundary.

There is a need to balance tangible landscape objectives for FLR with flexibility. Acknowledging that FLR is a process, and that identification of specific endpoints can be problematic under conditions of environmental, socio-economic and political uncertainty and/or instability, there remains a need for some guidance on direction towards an endpoint, if not a fully defined one. Multiple objectives may be set, for the short and long term, while recognizing that the process will require change to these over time. As landscapes are evolving systems, with multiple and complex influences, maintaining a focus on the desired direction can be delicate. Balancing this need for flexibility with the need for a clear direction can be tricky and

Figure 13.1 Understanding the multiple benefits that forests provide to rural communities in landscapes such as this one in western Maharashtra, India – including soil protection, stability of hydrological regimes, wood energy and biodiversity conservation – is an important prerequisite for balancing forest landscape restoration priorities among diverse actors.

Source: photo © John Parrotta.

lead to misconceptions among managers and landscape stakeholders. Indeed, ultimately, interventions in each landscape, over time, are likely to evolve quite differently, given the uniqueness of each context.

Landscape-level solutions should recognize diversity. There is no 'one size fits all' solution for FLR. Needs vary among stakeholders, and the ecological, historical, political, cultural and economic contexts are frequently a complex and unique mix. Stakeholders themselves evolve, and their needs may change over time. As a result, a regular reappraisal of the landscape context may be necessary, while a broad enough overarching set of objectives negotiated among stakeholders at the outset may better survive the test of time. Context-specific evaluations of FLR implications for ecological integrity and human wellbeing are needed. Unsustainable efforts have failed to recognize and understand this diversity and the attributes of the landscape that are of significance to people from other disciplines or sectors.

3 From the land sparing/land sharing debate and agroecological approaches we learned that...

Broad priorities for the landscape can integrate multiple objectives. Any landscape-level intervention – be it agriculture, forestry, mining or infrastructure development – should take place in tandem with other priorities for the landscape, based on stakeholders' identified needs and long-term objectives for the landscape. Understanding the widely varying motivations underlying choices by different stakeholders helps to build bridges across interest groups. Several restoration strategies can be implemented within a landscape approach to reconcile trade-offs between biodiversity conservation, agricultural production and human development. For example, a 2017 agreement between farmers and government authorities in the Putumayo region of Colombia (see Chapter 8) recognized the need to address deforestation caused by cattle ranching and to support locally adapted solutions, including development of silvopastoral systems, conservation and reforestation activities to be implemented by farmers, and the commercialization of non-timber forest products (NTFPs).

The human wellbeing dimension of FLR is multifaceted. Land sparing and land sharing emphasize the dichotomy between biodiversity conservation and food production. Moving beyond this dichotomy, human wellbeing – central to the FLR definition – can be measured through 'nature's contributions to people', which has pushed the land sparing/land sharing debate further. It builds on ecosystem services and the commodification of nature. In Chapter 7, the fine-level differences in the Paraiba watershed showed how different ecosystem services were improved by either land sparing or land sharing scenarios, demonstrating the need for a flexible approach to restoration measures.

Implementation may be constrained by a lack of measurable evidence. Large-scale measures and decision-support mechanisms such as the land sparing/land sharing concept generally lack sufficient quantifiable evidence to be applied systematically. Furthermore, contextual factors affect the choice of approach. Equally, there is a need to better quantify the benefits of restoration measures in landscape for people and biodiversity (or losses from degradation and forest loss) so as to better inform decisions. An understanding of the multiple potential objectives, and the processes necessary to achieve these objectives, helps to better inform decisions by stakeholders.

Agroecological approaches can reconcile divergent development and FLR visions. Innovative farming and agro-forest landscape management practices based on indigenous and local knowledge can contribute to the restoration of the productive, ecological, social and political fabric of landscapes degraded by conflict and/or socially and environmentally inappropriate agricultural development models that have ruptured these relationships. The case study of the Putumayo region in post-conflict Colombia demonstrated that integrated restoration efforts are in essence political, and touch upon contentious issues such as tenure and land rights, and diverging visions on development and conservation. The application of such approaches in FLR – provided that supportive government policies that empower farmers and their communities are in place – can help to resolve these tensions and support local practices that combine multiple objectives such as food production and biodiversity conservation.

Question 3: how can integrated approaches improve FLR decision-making processes?

Our final question sought to identify solutions to aid policymakers and other decision-makers interested in FLR.

1 Building on existing spatial planning approaches for FLR planning

A number of tools were presented, for example in Chapters 5, 6 and 7, demonstrating how to better visualize landscapes and how to integrate different stakeholders' views and perceptions. Modelling tools were illustrated in Chapter 6, such as, for example, engaging stakeholders in the Indonesian island of Boano to discuss historical events that have influenced their landscape, a first step towards engaging stakeholders in defining changes they would like to see in the future. Placing the emphasis on people, these tools are not as much about restoration techniques but, rather, about ways to bring different views and needs into the FLR equation. As this is a fundamental precursor to success in landscape-level interventions, more emphasis may be necessary to develop, disseminate and apply such tools. Many of these tools are simple but require different disciplinary training and, in many cases, interdisciplinary training. Investment in such tools and training could make a significant difference to the acceptability and local relevance of FLR and, therefore, to its implementation.

2 Broadening the knowledge base

Restoration projects in recent decades have emphasized Western science at the expense of indigenous and local knowledge, often resulting in misdiagnoses of forest/land degradation problems (e.g. impacts and benefits associated with shifting cultivation and other traditional agricultural practices) as well as disenfranchisement of key stakeholder groups. Indeed, Chapter 3 illustrated how Western knowledge systems are often pitted against Traditional knowledge rather than seeking to integrate the two as complementary ways of viewing the landscape. Yet, in many instances, as exemplified in this volume, the integration of both forms of knowledge can result in more comprehensive, locally grounded, and socially and culturally acceptable restoration projects. Integration of diverse forms of knowledge – including indigenous and local knowledge and practices, and associated decision-making models – into FLR planning and implementation can yield numerous benefits in terms of stakeholder engagement, conflict resolution, FLR implementation, monitoring and adaptive management, and on-the-ground results. This also entails better understanding stakeholders and their backgrounds and disciplines. Benefitting from different approaches can only prove an asset to FLR, with complementary knowledge enhancing the overall knowledge base of both ecological and social systems.

3 Increasing investment in the process of defining FLR objectives and targets

Short-term and top-down processes tend to dominate in FLR implementation. In reality, given the long-term nature of the process, upfront investment is necessary to better define agreed-upon FLR objectives and targets among stakeholders (at different spatial scales and considering different time scales). This includes assessment of current/potential benefits (for people and for ecological integrity) from the landscape under possible land use scenarios and specific management interventions. Participatory planning enables the incorporation of landscape stakeholders' needs and views in the overall process, offering major opportunities for multidisciplinary collaboration and for achieving multiple objectives. For example, in northern California, the Western Klamath Restoration Partnership (WKRP) described in Chapter 12 served to bring together tribal, federal and non-governmental stakeholders to define priority restoration actions for different parts of the landscape using multiple cultural values associated with the landscape and its use.

Figure 13.2 Integration of Western and Traditional knowledge in FLR planning and implementation can result in more comprehensive, locally grounded, and socially and culturally acceptable restoration projects. This scene from the Imlil Valley in the Atlas Mountains (Morocco) depicts a mosaic of agricultural and forest management practices used by farmers to support their livelihoods and food security while conserving forest biodiversity.

Source: photo © John Parrotta.

4 Adopting polycentric governance approaches in FLR planning and implementation to improve accountability, legitimacy and outcomes

Polycentricity and multi-level governance recognize diverse centres of decision-making, in line with the multiplicity of actors involved in FLR. Such governance arrangements encourage the participation of different actors in loose modes of collaboration across spatial scales and sectors. These modern forms of understanding and designing governance arrangements enable the inclusion of broader societal goals, leading to more sustainable approaches to landscape interventions such as FLR. Replacing top-down rules, policies and legislation with more flexible forms of governance is also a characteristic of polycentric approaches, which are more in line with the rapidly evolving dynamics of landscape work such as FLR. For example, as seen in Chapter 11, Habitat 141° in Australia, a diverse alliance that includes public, private and civil society actors, was created to allow stakeholders distributed across a large landscape to coordinate and collaborate on restoration activities.

5 Employing adaptive management that includes monitoring the effects of FLR implementation actions

If FLR is understood as an evolving and dynamic process, then defining clear and measurable indicators that can help assessors to verify progress towards objectives becomes more complex. Monitoring frameworks require clear and measurable environmental, economic and socio-cultural objectives as well as appropriate indicators that reflect progress towards these objectives. Equally, defining clear objectives becomes a negotiated process that evolves with time and will need to be regularly revisited. Keeping track of such flux in landscape objectives may require innovative means to identify progress, or, more importantly, lack of progress, and to take remedial actions. Even if specific endpoints are not easily defined and agreed upon at the outset, monitoring nonetheless reveals patterns and trends that can inform decision-making.

A final word ...

The contributions in this volume highlight the complexity surrounding the FLR process. They provide insight into the diverse tools and approaches that exist and can inform and guide FLR implementation. Integration signifies working across disciplines, sectors and spatial scales. This adds complexity. We need to find ways to incorporate this complexity and the knowledge that exists in other related areas into FLR implementation. Adaptive management is fundamental so that learning can inform future actions. We hope that through this book we may have provided a new lens on FLR implementation.

References

Aronson, J., Blignaut, J. N., Milton, S. J., Le Maitre, D., Esler, K. J., Limouzin, A., Fontaine, C., De Wit, M. P., Mugido, W., Prinsloo, P. and Van Der Elst, L. (2010) 'Are socioeconomic benefits of restoration adequately quantified? A meta-analysis of recent papers (2000–2008) in *Restoration Ecology* and 12 other scientific journals', *Restoration Ecology*, vol 18, no 2, pp. 143–154.

Brancalion, P. H. and Chazdon, R. L. (2017) 'Beyond hectares: four principles to guide reforestation in the context of tropical forest and landscape restoration', *Restoration Ecology*, vol 25, no 4, pp. 491–496.

Chazdon, R. and Guariguata, M. (2018) *Decision support tools for forest landscape restoration. Current status and future outlook*, CIFOR, Bogor.

Egan, D., Hjerpe, E. E. and Abrams, J., eds. (2011) *Human dimensions of ecological restoration: integrating science, nature, and culture*, Island Press, Washington, DC.

Emborg, J., Walker, G. and Daniels, S. (2012) Forest landscape restoration decision-making and conflict management: applying discourse-based approaches. In Stanturf, J., Lamb, D. and Madsen, P., eds, *Forest landscape restoration* (pp. 131–153), Springer, Dordrecht.

Keenleyside, K., Dudley, N., Cairns, S., Hall, C. and Stolton, S., 2012. *Ecological restoration for protected areas: principles, guidelines and best practice*, IUCN, Gland.

McGuire, D. (2014) 'FAO's forest and landscape restoration mechanism', *ETFRN News* 56: November 2014.

Newton, A. C., del Castillo, R. F., Echeverría, C., Geneletti, D., González-Espinosa, M., Malizia, L. R., Premoli, A. C., Rey Benayas, J. M., Smith-Ramírez, C. and Williams-Linera, G. (2012) 'Forest landscape restoration in the drylands of Latin America', *Ecology and Society*, vol 17, no 1, p. 21.

Pistorius, T. and Kiff, L. (2017) *From a biodiversity perspective: risks, tradeoffs, and international guidance for Forest Landscape Restoration*, UNIQUE, Freiburg.

Sterling, E. J., Filardi, C., Toomey, A., Sigouin, A., Betley, E., Gazit, N., Newell, J., Albert, S., Alvira, D., Bergamini, N. and Blair, M. (2017). 'Biocultural approaches to well-being and sustainability indicators across scales', *Nature Ecology & Evolution*, vol 1, no 12, p. 1798.

Vallauri, D., Aronson, J. and Dudley, N., 2005. An Attempt to Develop a Framework for Restoration Planning. In Mansourian, S., Vallauri, D. and Dudley, N., eds, *Forest restoration in landscapes: beyond planting trees* (pp. 65–72), Springer, New York.

Index

Page numbers in **bold** denote tables, those in *italics* denote figures.

For Product Safety Concerns and Information please contact our EU
representative GPSR@taylorandfrancis.com
Taylor & Francis Verlag GmbH, Kaufingerstraße 24, 80331 München, Germany